Statistics: A Very Short Introduction

VERY SHORT INTRODUCTIONS are for anyone wanting a stimulating and accessible way into a new subject. They are written by experts, and have been translated into more than 45 different languages.

The series began in 1995, and now covers a wide variety of topics in every discipline. The VSI library now contains over 500 volumes—a Very Short Introduction to everything from Psychology and Philosophy of Science to American History and Relativity—and continues to grow in every subject area.

Titles in the series include the following:

AFRICAN HISTORY John Parker and Richard Rathbone
AMERICAN HISTORY Paul S. Boyer
AMERICAN LEGAL HISTORY G. Edward White
AMERICAN POLITICAL PARTIES AND ELECTIONS L. Sandy Maisel
AMERICAN POLITICS Richard M. Valelly
AMERICAN SLAVERY Heather Andrea Williams
ANARCHISM Colin Ward
ANCIENT EGYPT Ian Shaw
ANCIENT GREECE Paul Cartledge
ANCIENT PHILOSOPHY Julia Annas
ANCIENT WARFARE Harry Sidebottom
ANGLICANISM Mark Chapman
THE ANGLO-SAXON AGE John Blair
ANIMAL RIGHTS David DeGrazia
ARCHAEOLOGY Paul Bahn
ARISTOTLE Jonathan Barnes
ART HISTORY Dana Arnold
ART THEORY Cynthia Freeland
ATHEISM Julian Baggini
THE ATMOSPHERE Paul I. Palmer
AUGUSTINE Henry Chadwick
BACTERIA Sebastian G. B. Amyes
BEAUTY Roger Scruton
THE BIBLE John Riches
BLACK HOLES Katherine Blundell
BLOOD Chris Cooper
THE BRAIN Michael O'Shea
THE BRICS Andrew F. Cooper
BRITISH POLITICS Anthony Wright
BUDDHA Michael Carrithers
BUDDHISM Damien Keown
BUDDHIST ETHICS Damien Keown
CAPITALISM James Fulcher
CATHOLICISM Gerald O'Collins
THE CELTS Barry Cunliffe
CHOICE THEORY Michael Allingham
CHRISTIANITY Linda Woodhead
CIRCADIAN RHYTHMS Russell Foster and Leon Kreitzman
CITIZENSHIP Richard Bellamy
CLASSICAL MYTHOLOGY Helen Morales
CLASSICS Mary Beard and John Henderson
CLIMATE CHANGE Mark Maslin
THE COLD WAR Robert McMahon
COMMUNISM Leslie Holmes
CONSCIOUSNESS Susan Blackmore
CONTEMPORARY ART Julian Stallabrass
COSMOLOGY Peter Coles
THE CRUSADES Christopher Tyerman
DADA AND SURREALISM David Hopkins
DARWIN Jonathan Howard
THE DEAD SEA SCROLLS Timothy Lim
DECOLONIZATION Dane Kennedy
DEMOCRACY Bernard Crick
DESIGN John Heskett
DREAMING J. Allan Hobson
DRUGS Les Iversen
THE EARTH Martin Redfern

ECONOMICS Partha Dasgupta
EGYPTIAN MYTH Geraldine Pinch
THE ELEMENTS Philip Ball
EMOTION Dylan Evans
EMPIRE Stephen Howe
ENGLISH LITERATURE Jonathan Bate
ETHICS Simon Blackburn
EUGENICS Philippa Levine
THE EUROPEAN UNION John Pinder and Simon Usherwood
EVOLUTION Brian and Deborah Charlesworth
EXISTENTIALISM Thomas Flynn
FASCISM Kevin Passmore
FEMINISM Margaret Walters
THE FIRST WORLD WAR Michael Howard
FORENSIC PSYCHOLOGY David Canter
FOUCAULT Gary Gutting
FREE SPEECH Nigel Warburton
FREE WILL Thomas Pink
FREUD Anthony Storr
FUNDAMENTALISM Malise Ruthven
FUNGI Nicholas P. Money
GALAXIES John Gribbin
GALILEO Stillman Drake
GAME THEORY Ken Binmore
GANDHI Bhikhu Parekh
GEOGRAPHY John Matthews and David Herbert
GEOPOLITICS Klaus Dodds
GLOBAL CATASTROPHES Bill McGuire
GLOBAL ECONOMIC HISTORY Robert C. Allen
GLOBALIZATION Manfred Steger
HABERMAS James Gordon Finlayson
HEGEL Peter Singer
HINDUISM Kim Knott
THE HISTORY OF LIFE Michael Benton
THE HISTORY OF MATHEMATICS Jacqueline Stedall
THE HISTORY OF MEDICINE William Bynum
THE HISTORY OF TIME Leofranc Holford-Strevens
HIV AND AIDS Alan Whiteside
HOLLYWOOD Peter Decherney
HUMAN EVOLUTION Bernard Wood
HUMAN RIGHTS Andrew Clapham
IDEOLOGY Michael Freeden
INDIAN PHILOSOPHY Sue Hamilton
INFINITY Ian Stewart
INFORMATION Luciano Floridi
INNOVATION Mark Dodgson and David Gann
INTELLIGENCE Ian J. Deary
INTERNATIONAL MIGRATION Khalid Koser
INTERNATIONAL RELATIONS Paul Wilkinson
ISLAM Malise Ruthven
ISLAMIC HISTORY Adam Silverstein
JESUS Richard Bauckham
JOURNALISM Ian Hargreaves
JUDAISM Norman Solomon
JUNG Anthony Stevens
KANT Roger Scruton
KNOWLEDGE Jennifer Nagel
THE KORAN Michael Cook
LAW Raymond Wacks
THE LAWS OF THERMODYNAMICS Peter Atkins
LEADERSHIP Keith Grint
LINGUISTICS Peter Matthews
LITERARY THEORY Jonathan Culler
LOCKE John Dunn
LOGIC Graham Priest
MACHIAVELLI Quentin Skinner
MARX Peter Singer
MATHEMATICS Timothy Gowers
THE MEANING OF LIFE Terry Eagleton
MEASUREMENT David Hand
MEDICAL ETHICS Tony Hope
MEDIEVAL BRITAIN John Gillingham and Ralph A. Griffiths
MEMORY Jonathan K. Foster
METAPHYSICS Stephen Mumford
MILITARY JUSTICE Eugene R. Fidell
MODERN ART David Cottington
MODERN CHINA Rana Mitter
MODERN IRELAND Senia Pašeta
MODERN ITALY Anna Cento Bull
MODERN JAPAN Christopher Goto-Jones
MODERNISM Christopher Butler
MOLECULES Philip Ball
MOONS David A. Rothery

MUSIC Nicholas Cook
MYTH Robert A. Segal
NEOLIBERALISM Manfred Steger and Ravi Roy
NIETZSCHE Michael Tanner
NORTHERN IRELAND Marc Mulholland
NOTHING Frank Close
THE PALESTINIAN-ISRAELI CONFLICT Martin Bunton
PANDEMICS Christian W. McMillen
PARTICLE PHYSICS Frank Close
PHILOSOPHY Edward Craig
PHILOSOPHY OF LAW Raymond Wacks
PHILOSOPHY OF SCIENCE Samir Okasha
PHOTOGRAPHY Steve Edwards
PLATO Julia Annas
POLITICAL PHILOSOPHY David Miller
POLITICS Kenneth Minogue
POSTCOLONIALISM Robert Young
POSTMODERNISM Christopher Butler
POSTSTRUCTURALISM Catherine Belsey
PREHISTORY Chris Gosden
PRESOCRATIC PHILOSOPHY Catherine Osborne
PSYCHIATRY Tom Burns
PSYCHOLOGY Gillian Butler and Freda McManus
QUANTUM THEORY John Polkinghorne
RACISM Ali Rattansi
THE REFORMATION Peter Marshall
RELATIVITY Russell Stannard
THE RENAISSANCE Jerry Brotton
RENAISSANCE ART Geraldine A. Johnson
REVOLUTIONS Jack A. Goldstone
RHETORIC Richard Toye
RISK Baruch Fischhoff and John Kadvany
RITUAL Barry Stephenson
RIVERS Nick Middleton
ROBOTICS Alan Winfield
ROMAN BRITAIN Peter Salway
THE ROMAN EMPIRE Christopher Kelly
THE ROMAN REPUBLIC David M. Gwynn
RUSSIAN HISTORY Geoffrey Hosking
THE RUSSIAN REVOLUTION S. A. Smith
SCHIZOPHRENIA Chris Frith and Eve Johnstone
SCIENCE AND RELIGION Thomas Dixon
SEXUALITY Véronique Mottier
SIKHISM Eleanor Nesbitt
SLEEP Steven W. Lockley and Russell G. Foster
SOCIAL AND CULTURAL ANTHROPOLOGY John Monaghan and Peter Just
SOCIALISM Michael Newman
SOCIOLOGY Steve Bruce
SOCRATES C. C. W. Taylor
SOUND Mike Goldsmith
THE SOVIET UNION Stephen Lovell
STATISTICS David J. Hand
STUART BRITAIN John Morrill
TELESCOPES Geoff Cottrell
THEOLOGY David F. Ford
TIBETAN BUDDHISM Matthew T. Kapstein
THE TUDORS John Guy
THE UNITED NATIONS Jussi M. Hanhimäki
THE VIKINGS Julian Richards
VIRUSES Dorothy H. Crawford
WAR AND TECHNOLOGY Alex Roland
WILLIAM SHAKESPEARE Stanley Wells
THE WORLD TRADE ORGANIZATION Amrita Narlikar

David J. Hand

STATISTICS

A Very Short Introduction

OXFORD
UNIVERSITY PRESS

Great Clarendon Street, Oxford OX2 6DP

Oxford University Press is a department of the University of Oxford.
It furthers the University's objective of excellence in research, scholarship,
and education by publishing worldwide in

Oxford New York

Auckland Cape Town Dar es Salaam Hong Kong Karachi
Kuala Lumpur Madrid Melbourne Mexico City Nairobi
New Delhi Shanghai Taipei Toronto

With offices in

Argentina Austria Brazil Chile Czech Republic France Greece
Guatemala Hungary Italy Japan Poland Portugal Singapore
South Korea Switzerland Thailand Turkey Ukraine Vietnam

Oxford is a registered trade mark of Oxford University Press
in the UK and in certain other countries

Published in the United States
by Oxford University Press Inc., New York

First Published 2008

British Library Cataloguing in Publication Data
Data available

Library of Congress Cataloging in Publication Data
Data available

ISBN 978-0-19-923356-4

Typeset by SPI Publisher Services, Pondicherry, India
Printed and bound by
CPI Group (UK) Ltd, Croydon, CR0 4YY

Contents

Preface

Statistical ideas and methods underlie just about every aspect of modern life. Sometimes the role of statistics is obvious, but often the statistical ideas and tools are hidden in the background. In either case, because of the ubiquity of statistical ideas, it is clearly extremely useful to have some understanding of them. The aim of this book is to provide such understanding.

Statistics suffers from an unfortunate but fundamental misconception which misleads people about its essential nature. This mistaken belief is that it requires extensive tedious arithmetic manipulation, and that, as a consequence, it is a dry and dusty discipline, devoid of imagination, creativity, or excitement. But this is a completely false image of the modern discipline of statistics. It is an image based on a perception dating from more than half a century ago. In particular, it entirely ignores the fact that the computer has transformed the discipline, changing it from one hinging around arithmetic to one based on the use of advanced software tools to probe data in a search for understanding and enlightenment. That is what the modern discipline is all about: the use of tools to aid perception and provide ways to shed light, routes to understanding, instruments for monitoring and guiding, and systems to assist decision-making. All of these, and more, are aspects of the modern discipline.

The aim of this book is to give the reader some understanding of this modern discipline. Now, clearly, in a book as short as this one, I cannot go into detail. Instead of detail, I have taken a high-level view, a bird's eye view, of the entire discipline, trying to convey the nature of statistical philosophy, ideas, tools, and methods. I hope the book will give the reader some understanding of how the modern discipline works, how important it is, and, indeed, why it is so important.

The first chapter presents some basic definitions, along with illustrations to convey some of the power, importance, and, indeed, excitement of statistics. The second chapter introduces some of the most elementary of statistical ideas, ideas which the reader may well have already encountered, concerned with basic summaries of data. Chapter 3 cautions us that the validity of any conclusions we draw depends critically on the quality of the raw data, and also describes strategies for efficient collection of data. If data provide one of the legs on which statistics stands, the other is probability, and Chapter 4 introduces basic concepts of probability. Proceeding from the two legs of data and probability, in Chapter 5 statistics starts to walk, with a description of how one draws conclusions and makes inferences from data. Chapter 6 presents a lightning overview of some important statistical methods, showing how they form part of an interconnected network of ideas and methods for extracting understanding from data. Finally, Chapter 7 looks at just some of the ways the computer has impacted the discipline.

I would like to thank Emily Kenway, Shelley Channon, Martin Crowder, and an anonymous reader for commenting on drafts of this book. Their comments have materially improved it, and helped to iron out obscurities in the explanations. Of course, any such which remain are entirely my own fault.

David J. Hand
Imperial College, London

List of illustrations

Chapter 1
Surrounded by statistics

> To those who say 'there are lies, damned lies, and statistics', I often quote Frederick Mosteller, who said that 'it is easy to lie with statistics, but easier to lie without them'.

Modern statistics

I want to begin with an assertion that many readers might find surprising: *statistics is the most exciting of disciplines*. My aim in this book is to show you that this assertion is true and to show you why it is true. I hope to dispel some of the old misconceptions of the nature of statistics, and to show what the modern discipline looks like, as well as to illustrate some of its awesome power, as well as its ubiquity.

In particular, in this introductory chapter I want to convey two things. The first is a flavour of the revolution that has taken place in the past few decades. I want to explain how statistics has been transformed from a dry Victorian discipline concerned with the manual manipulation of columns of numbers, to a highly sophisticated modern technology involving the use of the most advanced of software tools. I want to illustrate how today's statisticians use these tools to probe data in the search for structures and patterns, and how they use this technology to peel back the layers of mystification and obscurity, revealing the truths beneath. Modern statistics, like telescopes, microscopes, X-rays, radar, and medical scans, enables us to see things invisible to the naked eye. Modern statistics enables us to see through the mists and confusion of the world about us, to grasp the underlying reality.

So that is the first thing I want to convey in this chapter: the sheer power and excitement of the modern discipline, where it has come from, and what it can do. The second thing I hope to convey is the ubiquity of statistics. No aspect of modern life is untouched by it. Modern medicine is built on statistics: for example, the randomized controlled trial has been described as 'one of the simplest, most powerful, and revolutionary tools of research'. Understanding the processes by which plagues spread prevent them from decimating humanity. Effective government hinges on careful statistical analysis of data describing the economy and society: perhaps that is an argument for insisting that all those in government should take mandatory statistics courses. Farmers, food technologists, and supermarkets all implicitly use statistics to decide what to grow, how to process it, and how to package and distribute it. Hydrologists decide how high to build flood defences by analysing meteorological statistics. Engineers building computer systems use the statistics of reliability to ensure that they do not crash too often. Air traffic control systems are built on complex statistical models, working in real time. Although you may not recognize it, statistical ideas and tools are hidden in just about every aspect of modern life.

Some definitions

One good working definition of statistics might be that it is *the technology of extracting meaning from data*. However, no definition is perfect. In particular, this definition makes no reference to chance and probability, which are the mainstays of many applications of statistics. So another working definition might be that it is the *technology of handling uncertainty*. Yet other definitions, or more precise definitions, might put more emphasis on the roles that statistics plays. Thus we might say that statistics is the key discipline for *predicting the future* or for *making inferences about the unknown*, or for *producing convenient summaries of data*. Taken together these definitions broadly cover the essence of the discipline, though different applications will provide very different manifestations. For example, decision-making, forecasting, real-time monitoring, fraud detection, census enumeration, and analysis of gene sequences are all applications of statistics, and yet may require very different methods and tools. One thing to note about these definitions is that I have deliberately chosen the word 'technology' rather than science. A technology is the application of science and its discoveries, and that is what statistics is: the application of our understanding of how to extract information from data, and our understanding of uncertainty. Nevertheless, statistics is sometimes referred to as a science. Indeed, one of the most stimulating statistical journals is called just that: *Statistical Science*.

So far in this book, and in particular in the preceding paragraph, I have referred to the *discipline of statistics*, but the word 'statistics' also has another meaning: it is the plural of 'statistic'. A statistic is a numerical fact or summary. For example, a summary of the data describing some population: perhaps its size, the birth rate, or the crime rate. So in one sense this book is about individual numerical facts. But in a very real sense it is about much more than that. It is about how to collect, manipulate, analyse, and deduce things from those numerical facts. It is about the technology itself. This means that a reader hoping to find tables of numbers in this book (e.g. 'sports statistics') will be disappointed. But a reader hoping to gain understanding of how businesses make decisions, of how astronomers discover new types of stars, of how medical researchers identify the genes associated with a particular disease, of how banks decide whether or not to give someone a credit card, of how insurance companies decide on the cost of a premium, of how to construct spam filters which prevent obscene advertisements reaching your email inbox, and so on and on, will be rewarded.

All of this explains why 'statistics' can be both singular and plural: there is one discipline which is statistics, but there are many numbers which are statistics.

So much for the word 'statistics'. My first working definition also used the word 'data'. The word 'data' is the plural of the Latin word 'datum', meaning 'something given', from *dare*, meaning 'to give'. As such, one might imagine that it should be treated as a plural word: 'the data are poor' and 'these data show that...', rather than 'the data is poor' and 'this data shows that'. However, the English language changes over time. Increasingly, nowadays 'data' is treated as describing a continuum, as in 'the water is wet' rather than 'the water are wet'. My own inclination is to adopt whatever sounds more euphonious in any particular context. Usually, to my ears, this means sticking to the plural usage, but occasionally I may lapse.

Data are typically numbers: the results of measurements, counts, or other processes. We can think of such data as providing a simplified representation of whatever we are studying. If we are concerned with school children, and in particular their academic ability and suitability for different kinds of careers, we might choose to study the numbers giving their results in various tests and examinations. These numbers would provide an indication of their abilities and inclinations. Admittedly, the representation would not be perfect. A low score might simply indicate that someone was feeling ill during the examination. A missing value does not tell us much about their ability, but merely that they did not sit the examination. I will say more about *data quality* later. It matters because of the general principle (which applies throughout life, not merely in statistics) that if we have poor material to work with then the results will be poor. Statisticians can perform amazing feats in extracting understanding from numbers, but they cannot perform miracles.

Of course, many situations do not appear to produce numerical data directly. Much raw data appears to be in the form of pictures, words, or even things such as electronic or acoustic signals. Thus, satellite images of crops or rain forest coverage, verbal descriptions of side effects suffered when taking medication, and sounds uttered when speaking, do not appear to be numbers. However, close examination shows that, when these things are measured and recorded, they are translated into numerical representations or into representations which can themselves be further translated into numbers. Satellite pictures and other photographs, for example, are represented as millions of tiny elements, called pixels, each of which is described in terms of the (numerical) intensities of the different colours making it up. Text can be processed into word counts or measures of similarity between words and phrases; this is the sort of representation used by web search engines, such as Google. Spoken words are represented by the numerical intensities of the waveforms making up the individual parts of speech. In general, although not all data are numerical, most data are translated into numerical form at some stage. And most of statistics deals with numerical data.

Lies, damned lies, and setting the record straight

The remark that there are 'lies, damned lies, and statistics', which was quoted at the start of this chapter, has been variously attributed to Mark Twain and Benjamin Disraeli, among others. Several people have made similar remarks. Thus 'like dreams, statistics are a form of wish fulfilment' (Jean Baudrillard, in *Cool Memories*, Chapter 4); '... the worship of statistics has had the particularly unfortunate result of making the job of the plain, outright liar that much easier' (Tom Burnan, in *The Dictionary of Misinformation*, p. 246); 'statistics is "hocuspocus" with numbers'

(Audrey Habera and Richard Runyon, in *General Statistics*, p. 3); 'legal proceedings are like statistics. If you manipulate them, you can prove anything' (Arthur Hailey, in *Airport*, p. 385). And so on.

Clearly there is much suspicion of statistics. We might also wonder if there is an element of fear of the discipline. It is certainly true that the statistician often plays the role of someone who must exercise caution, possibly even being the bearer of bad news. Statisticians working in research environments, for example in medical schools or social contexts, may well have to explain that the data are inadequate to answer a particular question, or simply that the answer is not what the researcher wanted to hear. That may be unfortunate from the researcher's perspective, but it is a little unfair then to blame the statistical messenger.

In many cases, suspicion is generated by those who selectively choose statistics. If there is more than one way to summarize a set of data, all looking at slightly different aspects, then different people can choose to emphasize different summaries. A particular example is in crime statistics. In Britain, perhaps the most important source of crime statistics is the *British Crime Survey*. This estimates the level of crime by directly asking a sample of people of which crimes they have been victims over the past year. In contrast, the *Recorded Crime Statistics* series includes all offences notifiable to the Home Office which have been recorded by the police. By definition, this excludes certain minor offences. More importantly, of course, it excludes crimes which are not reported to the police in the first place. With such differences, it is no wonder that the figures can differ between the two sets of statistics, even to the extent that certain categories of crime may appear to be decreasing over time according to one set of figures but increasing according to the other.

The crime statistics figures also illustrate another potential cause of suspicion of statistics. When a particular measure is used as an indicator of the performance of a system, people may choose to target that measure, improving its value but at the cost of other aspects of the system. The chosen measure then improves disproportionately, and becomes useless as a measure of performance of the system. For example, the police could reduce the rate of shoplifting by focusing all their resources on it, at the cost of allowing other kinds of crime to rise. As a result, the rate of shoplifting becomes useless as an indicator of crime rate. This phenomenon has been termed 'Goodhart's law', named after Charles Goodhart, a former Chief Adviser to the Bank of England.

The point to all this is that the problem lies not with the statistics per se, but with the use made of those statistics, and the misunderstanding of how the statistics are produced and what they really mean. Perhaps it is perfectly natural to be suspicious of things we do not understand. The solution is to dispel that lack of understanding.

Yet another cause of suspicion arises in a fundamental way as a consequence of the very nature of scientific advance. Thus, one day we might read in the newspaper of a scientific study appearing to show that a particular kind of food is bad for us, and the next day that it is good. Naturally enough this generates confusion, the feeling that the scientists do not know the answer, and perhaps that they are not to be trusted. Inevitably, such scientific investigations make heavy use of statistical analyses, so some of this suspicion transfers to statistics. But it is the very essence of scientific advance that new discoveries are made that change our understanding. Where we once might have thought simply that dietary fat was bad for us, further studies may have led us to recognize that there are different kinds of fats, some beneficial and some detrimental. The picture is more complicated than we first thought, so it is hardly surprising that the initial studies led to conflicting and apparently contradictory conclusions.

A fourth cause of suspicion arises from elementary misunderstandings of basic statistics. As an exercise, the reader might try to decide what is suspicious about each of the following statements (the answers are in the endnote at the back of the book).

1) We read in a report that earlier diagnosis of a medical condition leads to longer survival times, so that screening programmes are beneficial.
2) We are told that a stated price has already been reduced by a 25% discount for eligible customers, but we are not eligible so we have to pay 25% more than the stated price.
3) We hear of a prediction that life expectancy will reach 150 years in the next century, based on simple extrapolation from increases over the past 100 years.
4) We are told that 'every year since 1950, the number of American children gunned down has doubled'.

Sometimes the misunderstandings are not so elementary, or, at least, they arise from relatively deep statistical concepts. It would be surprising if, after more than a century of development, there were not some deep counter-intuitive ideas in statistics. One such is known as the *Prosecutor's Fallacy*. It describes confusion between the probability that something will be true (e.g. the defendant is guilty) if you have some evidence (e.g. the defendant's gloves at the scene of the crime), with the probability of finding that evidence if you assume that the defendant is guilty. This is a common confusion, not merely in the courts, and we will examine it more closely later.

If there is suspicion and mistrust of statistics, it is clear that the blame lies not with the statistics or how they were calculated, but rather with the use made of those statistics. It is unfair to blame the discipline, or the statistician who extracts the meaning from the data. Rather, the blame lies with those who do not understand what the numbers are saying, or who wilfully misuse the results.

We do not blame a gun for murdering someone: rather it is the person firing the gun who is blamed.

Data

We have seen that data are the raw material on which the discipline of statistics is built, as well as the raw material from which individual statistics themselves are calculated, and that these data are typically numbers. In fact, however, data are more than merely numbers. To be useful, that is to enable us to carry out some meaningful statistical analysis, the numbers must be associated with some meaning. For example, we need to know what the measurements are measurements *of*, and just *what* has been counted when we are presented with a count. To produce valid and accurate results when we carry out our statistical analysis, we also need to know something about how the values have been obtained. Did everyone we asked give answers to a questionnaire, or did only some people answer? If only some answered, are they properly representative of the population of people we wish to make a statement about, or is the sample distorted in some way? Does, for example, our sample disproportionately exclude young people? Likewise, we need to know if patients dropped out of a clinical trial. And whether the data are up to date. We need to know if a measuring instrument is reliable, or if it has a maximum value which is recorded when the true value is excessively high. Can we assume that a pulse rate recorded by a nurse is accurate, or is it only a rough value? There is an infinite number of such questions which could be asked, and we need to be alert for any which could influence the conclusions we draw. Or else suspicions of the kind described above might be entirely legitimate.

One way of looking at data is to regard it as *evidence*. Without data, our ideas and theories about the world around us are mere speculations. Data provide a grounding, linking our ideas and theories to reality, and allowing us to validate and test our understanding. Statistical methods are then used to compare the data with our ideas and theories, to see how good a match there is. A poor match leads us to think again, to re-evaluate our ideas and reformulate them so that they better match what we actually observe to be the case. But perhaps I should insert a cautionary note here. This is that a poor match could also be a consequence of poor data quality. We must be alert for this possibility: our theories may be sound but our measuring instruments may be lacking in some way. In general, however, a good match between the observed data and what our theories say the data should be like reassures us that we are on the right track. It reassures us that our ideas really do reflect the truth of what is going on.

Implicit in this is that, to be meaningful, our ideas and theories must yield predictions, which we can compare with our data. If they do not tell us what we should expect to observe, or if the predictions are so general that any data will conform with our theories, then the theories are not much use: anything would do. Psychoanalysis and astrology have been criticized on such grounds.

Data also allow us to steer our way through a complex world – to make decisions about the best actions to take. We take our measurements, count our totals, and we use statistical methods to extract information from these data to describe how the world is behaving and what we should do to make it behave how we want. These principles are illustrated by aircraft autopilots, automobile SatNav systems, economic indicators such as inflation rate and GDP, monitoring patients in intensive care units, and evaluations of complex social policies.

Given the fundamental role of data as tying observations about the world around us to our ideas and understanding of that world, it is not stretching things too far to describe data, and the technology of extracting meaning from it, as the cornerstone of modern civilization. That is why I used the subtitle 'how data rule our world', for my book *Information Generation* (see Further reading).

Greater statistics

Although the roots can be traced as far back as we like, the discipline of statistics itself is really only a couple of centuries old. The Royal Statistical Society was established in 1834, and the American Statistical Association in 1839, whilst the world's first university statistics department was set up in 1911, at University College, London. Early statistics had several strands, which eventually combined to become the modern discipline. One of these strands was the understanding of probability, dating from the mid-17th century, which emerged in part from questions concerning gambling. Another was the appreciation that measurements are rarely error free, so that some analysis was needed to extract sensible meaning from them. In the early years, this was especially important in astronomy. Yet another strand was the gradual use of statistical data to enable governments to run their country. In fact, it is this usage which led to the word 'statistics': data about 'the State'. Every advanced country now has its own national statistical office.

As it developed, so the discipline of statistics went through several phases. The first, leading up to around the end of the 19th century, was characterized by discursive explorations of data. Then the first half of the 20th century saw the discipline becoming mathematicized, to the extent that many saw it as a branch of mathematics (it deals with numbers, doesn't it?). Indeed, university statisticians are still often based within mathematics departments. The second half of the 20th century saw the advent of the computer, and it was this change which elevated statistics from drudgery to excitement. The computer removed the need for practitioners to have special arithmetic skills – they no longer needed to spend endless hours on numerical manipulation. It is analogous to the change from having to walk everywhere to being able to drive: journeys which would have previously taken days now take a matter of minutes; journeys which would have been too lengthy to contemplate now become feasible.

The second half of the 20th century also saw the appearance of other schools of data analysis, with origins not in classical statistics but in other areas, especially computer science. These include machine learning, pattern recognition, and data mining. As these other disciplines developed, so there were sometimes tensions between the different schools and statistics. The truth is, however, that the varying perspectives provided by these different schools all have something to contribute to the analysis of data, to the extent that nowadays modern statisticians pick freely from the tools provided by all these areas. I will describe some of these tools later on. With this in mind, in this book I take a broad definition of statistics, following the definition of 'greater statistics' given by the eminent statistician John Chambers, who said: 'Greater statistics can be defined simply, if loosely, as everything related to *learning from data*, from the first planning or collection to the last presentation or report.' Trying to define boundaries between the different data-analytic disciplines is both pointless and futile.

So, modern statistics is not about calculation, it is about *investigation*. Some have even described statistics as the *scientific method in action*. Although, as I noted above, one still often finds many statisticians based in mathematics departments in universities, one also finds them in medical schools, social science departments, including economics, and many other departments, ranging from engineering to psychology. And outside universities large numbers work in government and in industry, in the pharmaceutical sector, marketing, telecoms, banking, and a host of other areas. All managers rely on statistical skills to help them interpret the data describing their department, corporation, production, personnel, etc. These people are not manipulating mathematical symbols and formulae, but are using statistical tools and methods to gain insight and understanding from evidence, from data. In doing so, they need to consider a wide variety of intrinsically non-mathematical issues such as data quality, how the data were or should be collected, defining the problem, identifying the broader objective of the analysis (understanding, prediction, decision, etc.), determining how much uncertainty is associated with the conclusion, and a host of other issues.

As I hope is clear from the above, statistics is ubiquitous, in that it is applied in all walks of life. This has had a reciprocal impact on the development of statistics itself. As statistical methods were applied in new areas, so the particular problems, requirements, and characteristics of those areas led to the development of new statistical methods and tools. And then, once they had been developed, these new methods and tools spread out, finding applications in other areas.

Some examples

Example 1: Spam filtering

'Spam' is the term used to describe unsolicited bulk email messages automatically sent out to many recipients, typically many millions of recipients. These messages will be advertising messages, often offensive, and they may be fronts for confidence tricksters. They include things such as debt consolidation offers, get-rich-quick schemes, prescription drugs, stock market tips, and dubious sexual aids. The principle underlying them is that if you email enough people, some are likely to be interested in – or taken in by – your offer. Unless the messages are from organizations specifically asked for information, most of them will be of no interest, and nobody will want to waste time reading and deleting them. Which brings us to spam filters. These are computer programs that automatically scan incoming email messages and decide which are likely to be spam. The filters can be set up so that the program deletes the spam messages automatically, sends them to a holding folder for later examination, or takes some other appropriate action. There are various estimates of the amount of spam sent out, but at the time of writing, one estimate is that over 90 billion spam messages are sent each day – and since this number has been rising dramatically month on month, it is likely to be substantially greater by the time you read this.

There are various techniques for preventing spam. Very simple approaches just check for the occurrence of keywords in the message. For example, if a message includes the word 'viagra' it might be blocked. However, one of the characteristics of spam detection is that it is something of an arms race. Once those responsible become aware that their messages are being blocked by a particular method, they seek ways round that method. For example, they might seek deliberately to misspell 'viagra' as 'v1agra' or 'v-iagra', so that you can recognize it but the automatic program cannot.

More sophisticated spam detection tools are based on statistical models of the word content of spam messages. For example, they might use estimates of the probabilities of particular words or word combinations arising in spam messages. Then a message that contains too many high-probability words is suspect. More sophisticated tools build models for the probability that one word will follow another, in a sequence, hence enabling the detection of suspicious phrases and sets of words. Yet other methods use statistical models of images, to detect such things as skin tones in an emailed picture.

Example 2: The Sally Clark case

In 1999, Sally Clark, a young British lawyer, was tried, convicted, and given a life sentence for murdering her two baby sons. Her first child died in 1996, aged 11 weeks, and her second died in 1998, aged 8 weeks. The verdict depended on what has become a byword for the misunderstanding and misuse of statistics, when the paediatrician Sir Roy Meadow, in his role as expert witness for the prosecution, claimed that the chance of two children dying from cot death was 1 in 73 million. He obtained this figure by simply multiplying together the chance for the two deaths separately. In doing so, in his ignorance of basic statistics, he entirely ignored the fact that one such death in a family is likely to mean that another such death is more likely.

Study of past data shows that the probability of a randomly selected baby suffering a cot death in a family such as the Clarks' is about 1 in 8,500. If one then makes the assumption that the occurrence of one such death does not change the probability of another, then the chances of two such deaths in the same family would be 1/8,500 times 1/8,500; that is, about one in 73 million. But the assumption here is a big one, and careful statistical analysis of past data suggests that, in fact, the chance of a second cot death is substantially increased when one has already occurred. Indeed, the calculations suggest that several such multiple deaths should be expected to occur each year in a nation the size of the UK. The website of the Foundation for the Study of Infant Death says 'it is very rare for cot death to occur twice in the same family, though occasionally an inherited disorder, such as a metabolic defect, may cause more than one infant to die unexpectedly'.

In the Sally Clark case, there was more evidence suggesting that she was innocent, and eventually it became clear that her second son had a bacterial infection known to predispose towards sudden infant death. Ms Clark was subsequently released on appeal in 2003. Tragically, she died in March 2007, aged just 42. More details of this terrible misunderstanding and misuse of statistics are given in an excellent article by Helen Joyce and on the website listed in the Further reading at the end of this book.

Example 3: Star clusters

As our ability to probe further and further into the universe has increased, so it has become apparent that astronomic objects tend to cluster together, and do so in a hierarchical way, so that stars form clusters, clusters of stars themselves form higher level clusters, and these then cluster in turn. In particular, our own galaxy, which is a cluster of stars, is a member of the *Local Group* of about thirty galaxies, and this in turn is a member of the *Local Supercluster*. At the largest scale, the Universe looks rather like a foam, with filaments consisting of Superclusters lying on the edges of vast empty spaces. But how was all this discovered? Even if we use powerful telescopes to look out from the Earth, we simply see a sky of stars. The answer is that teasing out this clustering structure, and indeed discovering it in the first place, required statistical techniques. One class of techniques involves calculating the distances from each star to its few closest stars. Stars which have more stars closer than expected by chance are in locally dense regions – local clusters.

Of course, there is much more to it than that. Interstellar dust clouds will obscure the view of distant objects, and these dust clouds are not distributed uniformly in space. Likewise, faint objects will only be seen if they are near enough to the Earth. A thin filament of galaxies seen end on from the Earth could appear to be a dense cluster. And so on. Sophisticated statistical corrections need to be applied so that we can discern the underlying truth from the apparent distributions of objects.

Understanding the structure of the universe sheds light both on how it came to be, and on its future development.

Example 4: Manufacturing chemicals

I have already remarked that while statisticians may be able to perform amazing feats, they cannot perform miracles. In particular, the quality of their conclusions will be moderated by the quality of the data. Given this, it is hardly surprising that there are important subdisciplines of statistics concerned with how best to collect data. These are discussed in Chapter 3. One of these subdisciplines is *experimental design*. Experimental design techniques are used in situations where it is possible to control or manipulate some of the 'variables' being studied. The tools of experimental design enable us to extract maximum information for a given use of resources. For example, in producing a particular chemical polymer we might be able to set the temperature, pressure, and time of the chemical reaction to any values we want. Different values of these three variables will lead to variations in the quality of the final product. The question is, what is the best set of values?

In principle, this is an easy question to answer. We simply make many batches of the polymer, each with different values of the three variables. This allows us to estimate the 'response surface', showing the quality of the polymer at each set of three values of the variables, and we can then choose the particular triple which maximizes the quality.

But what if the manufacturing process is such that it takes several days to make each batch? Making many such batches, just to work out the best way of doing so, may be infeasible. Making 100 batches, each of which takes three days, would take the better part of a year. Fortunately, cleverly designed experiments allow us to extract the same information from far fewer carefully chosen sets of values. Sometimes a tiny fraction of batches can yield enough information for us to determine the best set of values, provided those batches are properly selected.

Example 5: Customer satisfaction

To run any retail organization effectively, so that it makes a profit and grows over time, requires paying careful attention to the customers, and giving them the product or service that they want. Failing to do so will mean that they go to a competitor who does provide what is wanted. The bottom line here is that failure will be indicated by declining revenues. We can try to avoid that by collecting data on how the customers feel before they begin voting with their wallets. We can carry out surveys of customer satisfaction, asking customers if they are happy with the product or service and in what ways these might be improved.

At first glance, it might look as if, to obtain reliable conclusions which reflect the behaviour of the entire customer base, it is necessary to give questionnaires to all the customers. This could clearly be an expensive and time-consuming exercise. Fortunately, however, there are statistical methods which enable sufficiently accurate results to be obtained from just a sample of customers. Indeed, the results can sometimes be even more accurate than surveying all customers. Needless to say, great care is needed in such an exercise. It is necessary to be wary of basing conclusions on a distorted sample: the results would be useless as a description of how customers behaved in general if only those who spent large sums of money were interviewed. Once again, statistical methods have been developed which enable us to avoid such mistakes – and so to draw valid conclusions.

Example 6: Detecting credit card fraud

Not all credit card transactions are legitimate. Fraudulent transactions cost the bank money, and also cost the bank's customers money. Detecting and preventing fraud is thus very important. Many readers of this book will have had the experience of their bank telephoning them to check that they made certain transactions. These calls are based on the predictions made by statistical models which describe how legitimate customers behave. Departures from the behaviour predicted by these models suggest that something suspicious is going on, deserving investigation.

There are various kinds of model. Some are based simply on intrinsically suspicious patterns of behaviour: simultaneous use of a single card in geographically distant locations, for example.

Others are based on more elaborate models of the kinds of transactions someone habitually makes, when they tend to make them, for how much money, at what kinds of outlets, for which kinds of products, and so on.

Of course, no such predictive model is perfect. Credit card transactions patterns are often varied, with people suddenly making purchases of a kind they have never made before. Moreover, only a tiny percentage of transactions are fraudulent – perhaps around one in a thousand. This makes detection especially difficult.

Detecting and preventing fraud is a constant battle: when one fraud avenue is stopped, fraudsters tend not to abandon their chosen career path and take up a legitimate occupation, but switch to other methods of fraud, so requiring the development of further statistical models.

Example 7: Inflation

We are all familiar with the notion that things become more expensive as time passes. But how can we compare today's cost of living with yesterday's? To do so, we need to compare the same things bought on the two dates. Unfortunately, there are complications: different shops charge different prices for the same things, different people buy different things, the same people change in their purchasing patterns, new products appear on the market and old ones vanish, and so on. How can we allow for changes such as these in determining whether life is more expensive nowadays?

Statisticians and economists construct indicators such as the Retail Price Index and the Consumer Price Index to measure the cost of living. These are based on a notional 'basket' of (hundreds of) goods that people buy, along with surveys to discover what prices are being charged for each item in the basket. Sophisticated statistical models are used to combine the prices of the different items to yield a single overall number which can be compared over time. As well as serving as an indicator of inflation, such indices are also used to adjust tax thresholds and index-linked salaries, pensions, and so on.

Conclusion

It may not always be apparent to the untutored eye, but statistics and statistical methods lie at the heart of scientific discovery, commercial operations, government, social policy, manufacturing, medicine, and most other aspects of human endeavour. Furthermore, as the world progresses, so this role is becoming more and more important. For example, the development of new medicines has long had a legal requirement for statisticians to be involved and something similar is now happening in the banking industry, with new international agreements requiring statistical risk models to be built. Given this pivotal role, it is clearly important that no educated citizen should be unaware of basic statistical principles.

Modern statistics, with its use of sophisticated software tools to probe data, permits us to make voyages of discovery paralleling those of pre-20th-century explorers, investigating new and exciting realms. This recognition – that real statistics is about exploring the unknown, not about tedious arithmetic manipulation – is central to an appreciation of the modern discipline.

Chapter 2
Simple descriptions

Data are nature's evidence

Introduction

In this chapter, I aim to introduce some of the basic concepts and tools which form the foundation of statistics, and which enable it to play its many roles.

In Chapter 1, I noted that modern statistics suffered from many misconceptions and misunderstandings. Yet another such misunderstanding is often (probably inadvertently) propagated by textbooks which describe statistical methods for experts in other disciplines. This is that statistics is a bag of tools, with the role of the statistician or user of statistics being to pick one tool to match the question, and then to apply it.

The problem with this view of statistics is that it gives the impression that the discipline is simply a collection of disconnected methods of manipulating numbers. It fails to convey the truth that statistics is a connected whole, built on deep philosophical principles, so that the data analytic tools are linked and related: some may generalize to others, some may appear to differ simply because they work with different kinds of data, even though they search for the same kind of structures, and so on. I suspect that this impression of a collection of isolated methods may be another reason why newcomers find statistics rather tedious and hard to learn (apart from any fear of numbers they may have). Learning a disconnected and apparently quite distinct set of methods is much tougher than learning about such methods through their relationship of derivation from underlying principles. It is rather like the difficulty of learning a random collection of unrelated words, compared with learning words in a meaningful sentence. I have endeavoured, in this chapter and throughout the book, to convey the relationships between statistical ideas, to show that the discipline is really an interconnected whole.

Data again

Whatever else it does, and whatever the details of the definition we adopt, statistics begins with data. Data describe the universe we wish to study. I am using the word 'universe' here in a very general sense. It could be the physical world about us, but it could be the world of credit card transactions, of microarray experiments in genetics, of schools and their teaching and examination performance, of trade between countries, of how people behave when exposed to different advertisements, of subatomic particles, and so on. There is no end to the worlds which can be studied, and therefore of the worlds represented by data.

Of course, no finite data set can tell us about all of the infinite complexities of the real world, just as no verbal description, even that written by the most eloquent of authors, can convey everything about every facet of the world around us. That means we must be specially aware of any potential shortcomings or gaps in our data. It means that, when collecting data, we need to take special care to ensure that they do cover the aspects we are interested in, or about which we wish to draw conclusions. There is also a more positive way of looking at this: by collecting only a finite set of descriptive aspects, we are forced to eliminate the irrelevant ones. When studying the safety of different designs of cars, we might decide not to record the colour of the fabric covering the seats.

Broadly speaking, it is convenient to regard data as having two aspects. One aspect is concerned with the objects we wish to study, and the other aspect is concerned with the characteristics of those objects we wish to study. For example, our objects might be children at school and the characteristics might be their test scores. Or perhaps the objects might be children, but we are studying their diet and physical development, in which case the characteristics might be the children's height and weight. Or our objects might be physical materials, with the characteristics of interest being their electrical and magnetic properties. In statistics, it is common to call the characteristics *variables*, with each object having a *value* of a variable (the child's score in a spelling test would be the value of the test variable, the magnitude of material's electrical conductivity would be the value of the conductivity variable, etc.). In other data-analytic disciplines, alternative words are sometimes used (such as 'feature', 'characteristic', or 'attribute'), but when I get on to discussing the technical aspects I shall usually stick to 'variable'.

In fact, in any one study we might be interested in multiple kinds of objects. We might want to understand and make statements not only about school children, but also about the schools themselves, and perhaps about the teachers, the styles of teaching, and different kinds of school management structure, all in one study. Moreover, we will typically not be interested in any single characteristic of the objects being studied, but in relationships between characteristics, and indeed, perhaps relationships between characteristics for objects of different kinds and at different levels. We see that things are often really quite complicated, as we might expect, given the complexity of the subjects we might be studying.

Many people are resistant to the notion that numerical data can convey the beauty of the real world. They feel that somehow converting things to numbers strips away the magic. In fact, they could not be more wrong. Numbers have the potential to allow us to perceive that beauty, that magic, more clearly and more deeply, and to appreciate it more fully. Admittedly, *ambiguity* may be removed by couching things in numerical form: if I say that there are four people in the room, you know exactly what I mean, whereas, in contrast, if I say that someone is attractive you may not be entirely sure what I mean. You may even disagree with my view that someone is attractive, but you are unlikely to disagree with my view that there are four people in the room (barring errors in our counting, of course, but that's a different matter). Numbers are universally understood, regardless of nationality, religion, gender, age, or any other human characteristic. Removing ambiguity, and with it removing the risk of misunderstanding, can only be beneficial when trying to understand something – when trying to see to its heart.

This lack of ambiguity in the interpretation of numbers is closely tied to the fact that *numbers have only one property*: their value or magnitude. Contrary to what fortune tellers may have us believe, numbers are not lucky or unlucky – in just the same way that numbers do not have a colour, or a flavour, or an odour. They have no properties but their intrinsic numerical value. (Admittedly, some people experience *synaesthesia*, in which they do associate a particular colour or sensation with particular numbers. However, the associated sensations are different for different people, and cannot be regarded as properties of the numbers themselves.)

Numerical data give us a more direct and immediate link to the phenomena we are studying than do words, because numerical data are typically produced by measuring instruments with a more direct link to those phenomena than are words. Numbers come directly from the things being studied, whereas words are filtered by a human brain. Of course, things are more complicated if our data-collection procedure is mediated by words (as would be the case if the data are collected by questionnaires), but the principle still holds good. While measuring instruments may not be perfect, the data are a proper representation of the results of applying those instruments to the phenomenon being investigated. I sometimes summarize this by the comment at the start of this chapter: *data are nature's evidence, seen through the lens of the measuring instrument.*

On top of all this, numbers have practical consequences in terms of societal advance. It is the civilized world's facility with manipulating the representations of reality provided by numbers that has led to such awesome material progress in the past few centuries.

Although numbers have only one property, their numerical value, we might choose to use that property in different ways. For example, when deciding on the order of merit of students in a class, we might rank them according to their examination scores. That is, we might care only about whether one score is higher than another, and not about the precise numerical difference. When we are concerned only with the *order* of the values in this way we say we are treating the data as lying on an 'ordinal' scale. On the other hand, when a farmer measures the amount of corn he has produced, he does not simply want to know whether he has grown more than he grew last year. He also wants to know how much he has produced: its actual weight. It is on this basis, after all, that it will be sold in the market. In this situation, the farmer is really comparing the weight of corn he has produced with a standard weight, such as a ton, so that he can say how many tons of corn he has produced. Implicit in this is the calculation of the ratio of the weight of the corn the farmer has produced to the weight of one ton of corn. For this reason, when we use the values in this way, we say we are treating the data as lying on a 'ratio' scale. Note that in this case we could choose to change the basic unit of measurement: we could calculate the weight in pounds or kilograms rather than tons. As long as we say what unit we have used, then it is easy for anyone else to convert back, or to convert to whatever unit they normally use.

In yet another situation, we might want to know how many patients have suffered from a particular side effect of a medicine. If the number is large enough we might want to withdraw the drug from the market as being too risky. In this case, we are simply counting discrete well-defined units (patients). No rescaling by changing units would be meaningful (we would not contemplate counting the number of 'half patients'!), so we say we are treating the data as lying on an 'absolute' scale.

Simple summary statistics

Whilst simple numbers constitute the *elements* of data, in order for them to be useful we need to look at the relationships between them, and perhaps combine them in some way. And this is where statistics comes in. Later chapters will explore more complex ways of comparing and combining numbers, but this chapter serves to introduce the ideas. Here we look at some of the most straightforward ways: we will not explore relationships between different variables in this chapter, but simply look at information and insights which can be extracted from relationships between values measured on the same variable. For example, we might have recorded the ages of the applicants for a place at a university, the luminosity of the stars in a cluster, the monthly expenditures of families in a town, the weights of cows in a herd at the time of sending them to market, and so on. In each case, a single numerical value is recorded for each 'object' in a population of objects.

The individual values in the collection, when taken together, are said to form a 'distribution' of values. Summary statistics are ways of characterizing that distribution: of saying whether the values are very similar, whether there are some exceptionally large or small values, what a 'typical' value is like, and so on.

Averages

One of the most basic kinds of descriptions, or summary statistics, of a set of numbers is an 'average'. An average is a representative value; it is close, in some sense, to the numbers in the set. The need for such a thing is most apparent when the set of numbers is large. For example, suppose we had a table recording the ages of each of the people in a large city – perhaps with a million inhabitants. For administrative and business purposes it would obviously be useful to know the average age of the inhabitants. Very different services would be needed and sales opportunities would arise if the average age was 16 instead of 60. We could try to get a ball-park feel for the general size of the numbers in the table, the ages, by looking at each of the values. But this would clearly be a tough exercise. Indeed, if it took only one second to look at each number, it would take over 270 hours to look through a table of a million numbers, and that's ignoring the actual business of trying to remember and compare them. But we can use our computer to help us.

First, we need to be clear about exactly what we mean by 'average', because the word has several meanings. Perhaps the most widely used type of average is the *arithmetic mean*, or just *mean* for short. If people use the word 'average' without saying how they interpret it, then they probably intend the arithmetic mean.

Before I show how to calculate the arithmetic mean, imagine another table of a million numbers. Only, in this second table, suppose that all the numbers are identical to each other. That is, suppose that they all have the same value. Now add up all the numbers in the first table, to find their total (this takes but a split second using a computer). And add up all the numbers in the second table, to find their total. If the two totals are the same, then the number which is repeated a million times in the second table is capturing some sort of essence of the numbers in the first table. This single number, for which a million copies add up to the same total as the first table, is called the arithmetic mean (of the numbers in the first table).

In fact, the arithmetic mean is most easily calculated simply by dividing the total of the million numbers in the first table by a million. In general, the arithmetic mean of a set of numbers is found by adding all the numbers up and dividing by how many there are. Here is a further example. In a test, the percentage scores for five students in a class were 78, 63, 53, 91, and 55. The total is $78 + 63 + 53 + 91 + 55 = 340$. The arithmetic mean is then simply given by dividing 340 by 5. It is 68. We would get the same total of 340 if all five students each scored the mean value, 68.

The arithmetic mean has many attractive properties. It always takes a value between the largest and smallest values in the set of numbers. Moreover, it balances the numbers in the set, in the sense that the sum of the differences between the arithmetic mean and those values larger than it is exactly equal to the sum of the differences between the arithmetic mean and those values smaller than it. In that sense, it is a 'central' value. Those of a mechanical turn of mind might like to picture a set of 1kg weights placed at various positions along a (weightless) plank of wood. The distances of the weights from one end of the plank represent the values in the set of numbers. The mean is the distance from the end such that a pivot placed there would perfectly balance the plank.

The arithmetic mean is a *statistic*. It summarizes the entire set of values in our collection to a single value. It follows from this that it also throws away information: we should not expect to represent a million (or five, or however many) different numbers by a single number without sacrificing something. We shall explore this sacrifice later. But since it is a central value in the sense illustrated above it can be a useful summary. We can compare the average class size in different schools, the average test score of different students, the average time it takes different people to get to work, the average daily temperature in different years, and so on.

The arithmetic mean is one important statistic, a summary of a set of numbers. Another important summary is the *median*. The mean was the pivotal value, a sort of central point balancing the sum of differences between it and the numbers in the set. The median balances the set in another way: it is the value such that half the numbers in the data set are larger and half are smaller. Returning to the class of five students illustrated above, their scores, in order from smallest to largest, are 53, 55, 63, 78, and 91. The middle score here is 63, so this is the median.

Obviously complications arise if there are equal values in the data set (e.g. suppose it consists of 99 copies of the value 0 and a single copy of the value 1), but these can be overcome. In any case, once again the median is a representative value in some sense, although in a different sense from the mean. Because of this difference, we should expect it to take a value different from the mean. Obviously the median is easier to calculate than the mean. We do not even have to add up any values to reach it, let alone divide by the number of values in the set. All we have to do is order the numbers, and locate the one in the middle. But in fact this computational advantage is essentially irrelevant in the computer age: in real statistical analyses the computer takes over the tedium of arithmetic juggling.

Presented with these two summary statistics, both providing representative values, how should we choose which to use in any particular situation? Since they are defined in different ways, combining the numerical values differently, they are likely to produce different values, so any conclusions based on them may well be different. A full answer to the question of which to choose would get us into technicalities beyond the level of this book, but a short answer is that the choice will depend on the precise details of the question one wishes to answer.

Here is an illustration. Suppose that a small company has five staff, each in a different grade and earning, respectively, \$10,000, \$10,001, \$10,002, \$10,003, and \$99,999. The mean of these is \$28,001, while the median is \$10,002. Now suppose that the company intends to recruit five new employees, one to each grade. The employer might argue that in this case, 'on average', she would have to pay the newcomers a salary of \$28,001, so that this is the average salary she states in the advertisement. The employees, however, might feel that this is dishonest, since as many new employees will be paid less than \$10,002 as will be paid more than \$10,002. They might feel it is more honest to put this figure in the advertisement. Sometimes it requires careful thought to decide which measure is appropriate. (And in case you think this argument is contrived, Figure 1 shows the distribution of American baseball players' salaries prior to the 1994 strike. The arithmetic mean was \$1.2 million, but the median was only \$0.5 million.)

This example also illustrates the relative impact of extreme values on the mean and the median. In the pay example above, the mean is nearly three times the median. But suppose the largest value had been \$10,004 instead of \$99,999. Then the median would remain as \$10,002 (half the values above and half below), but the mean would shrink to \$10,002. The size of just a single value can have a dramatic effect on the mean, but leave the median untouched. This sensitivity of the mean to extreme values is one reason why the median may sometimes be chosen in preference to the mean.

The mean and the median are not the only two representative value summaries. Another important one is the *mode*. This is the value taken most frequently in a sample. For example, suppose that I count the number of children per family for families in a certain population. I might find that some families have one child, some two, some three, and so on, and, in particular, I might find that more families have two children than any other value. In this case, the mode of the number of children per family would be two.

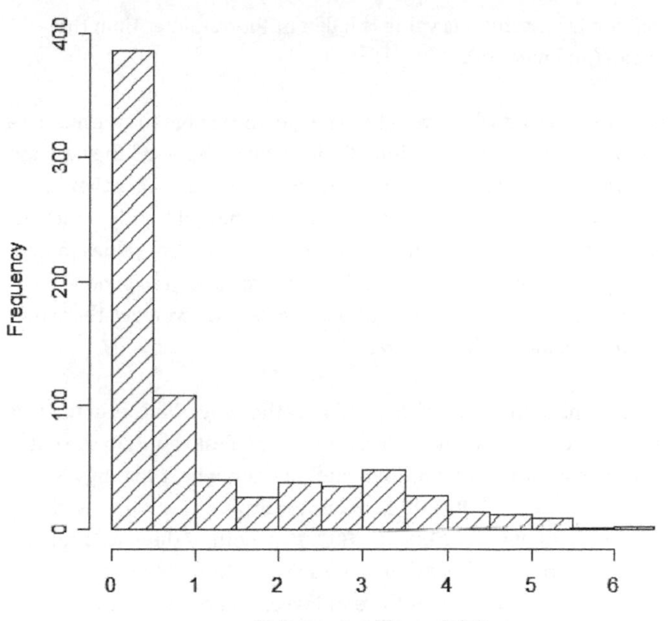

1. Distribution of American baseball players' salaries in 1994. The horizontal axis shows salaries in millions of dollars, and the vertical axis the numbers in each salary range

Dispersion

Averages, such as the mean and the median, provide single numerical summaries of collections of numerical values. They are useful because they can give an indication of the general size of the values in the data. But, as we have seen in the example above, single summary values can be misleading. In particular, single values might deviate substantially from individual values in a set of numbers. To illustrate, suppose that we have a set of a million and one numbers, taking the values 0, 1, 2, 3, 4, . . . , 1,000,000. Both the mean and the median of this set of values are 500,000. But it is readily apparent that this is not a very 'representative' value of the set. At the extremes, one value in the set is half a million larger and one value is half a million smaller than the mean (and median).

What is missing when we rely solely on an average to summarize a set of data is some indication of how widely dispersed the data are around that average. Are some data points much larger than the average? Are some much smaller? Or are they all tightly bunched about the average? In general, how different are the values in the data set from each other? Statistical measures of *dispersion* provide precisely this information, and as with averages there is more than one such measure.

The simplest measure of dispersion is the *range*. This is defined as the difference between the largest and smallest values in the data set. In our data set of a million and one numbers, the range is 1,000,000 – 0 = 1,000,000. In our example of five salaries, the range is $99,999 – $10,000 = $89,999. Both of these examples, with large ranges, show that there are substantial departures from thc mean. For example, if the employees had been earning the respective salaries of $27,999, $28,000, $28,001, $28,002, $28,003 then the mean would also be $28,001, but the range would be only $4. This paints a very different picture, telling us that the employees with these new salaries earn much the same as each other. The large range of the earlier example, $89,999, immediately tells us that there are gross differences.

The range is all very well, and has many attractive properties as a measure of dispersion, not least its simplicity and ready interpretability. However, we might feel that it is not ideal. After all, it ignores most of the data, being based on only the largest and smallest values. To illustrate, consider two data sets, each consisting of a thousand values. One data set has one value of 0, 998 values of 500, and one value of 1000. The other data set has 500 values of 0 and 500 values of 1000. Both of these data sets have a range of 1000 (and, incidentally, both also have a mean of

500), but they are clearly very different in character. By focusing solely on the largest and smallest values, the range has failed to detect the fact that the first data set is mostly densely concentrated around the mean.

This shortcoming can be overcome by using a measure of dispersion which takes *all* of the values into account.

One common way to do this is to take the differences between the (arithmetic) mean and each number in the data set, square these differences, and then find the mean of these squared differences. (Squaring the differences makes the values all positive, otherwise positive and negative differences would cancel out when we calculated the mean.) If the resulting mean of the squared differences is small, it tells us that, on average, the numbers are not too different from their mean. That is, they are not widely dispersed. This mean squared difference measure is called the *variance* of the data – or, in some disciplines, simply the *mean squared deviation*. Illustrating with our five students, their test scores were 78, 63, 53, 91, and 55 and their mean was 68. The squared differences between the first score and the mean is $(78 - 68)^2 = 100$, and so on. The sum of the squared differences is $100 + 25 + 225 + 529 + 169 = 1048$, so that the mean of the squared differences is $1048 \div 5 = 209.6$. This is the variance.

One slight complication arises from the fact that the variance involves squared values. This implies that the variance itself is measured in 'square units'. If we measure the productivity of farms in terms of tons of corn, the variance of the values is measured in 'tons squared'. It is not obvious what to make of this. Because of this difficulty, it is common to take the square root of the variance. This changes the units back to the original units, and produces the measure of dispersion called the *standard deviation*. In the example above, the standard deviation of the students' test scores is the square root of 209.6, or 14.5.

The standard deviation overcomes the problem that we identified with the range: it uses all of the data. If most of the data points are clustered very closely together, with just a few outlying points, this will be recognized by the standard deviation being small. In contrast, if the data points take very different values, even if they have the same largest and smallest value, the standard deviation will be much larger.

Skewness

Measures of dispersion tell us how much the individual values deviate from each other. But they do not tell us in what way they deviate. In particular, they do not tell us if the larger deviations tend to be for the larger values or the smaller values in the data set. Recall our example of the five company employees, in which four earned about $10,000 per year, and one earned around ten times that. A measure of dispersion (the standard deviation, for example) would tell us that the values were quite widely spread out, but would not tell us that one of the values was much larger than the others. Indeed, the standard deviation for the five values $90,000, $89,999, $89,998, $89,997, and $1 is exactly the same as for the original five values. What is different is that the anomalous value (the $1 value) is now very small instead of very large. To detect this difference, we need another statistic to summarize the data, one which picks up on and measures the *asymmetry* in the distribution of values. One kind of asymmetry in distributions of values is called *skewness*. Our original employee salary example, with one anomalously large value of $99,999, is *right skewed* because the distribution of values has a long 'tail' stretching out to the single very large value of $99,999. This distribution has many smaller values and very few larger values. In contrast, the distribution of values given above, in which $1 is the anomaly, is *left skewed*, because the bulk of the values bunch together and there is a long tail stretching down to the single very small value.

Right skewed distributions are very common. A classic example is the distribution of wealth, in which there are many individuals with small sums and just a few individuals with many billions of dollars. The distribution of baseball players' salaries in Figure 1 is heavily right skewed.

Quantiles

Averages, measures of dispersion, and measures of skewness provide overall summary statistics, condensing the values in a distribution down to a few convenient numbers. We might, however, be interested in just parts of a distribution. For example, we might be concerned with just the largest or smallest few – say, the largest 5% – values in the data set. We have already met the median, the value which is in the middle of the data in the sense that 50% of the values are larger and 50% are smaller. This idea can be generalized. For example, the *upper quartile* of a set of numbers is that value such that 25% (i.e. a quarter) of the data values are larger, and the *lower quartile* is that value such that 25% of the data values are smaller.

This is taken further to produce *deciles* (dividing the data set into tenths, from the lowest tenth through to the highest tenth) and *percentiles* (dividing the data into 100ths). Thus someone might be described as scoring above the 95th percentile, meaning that they are in the top 5% of the set of scores. The general term, including quartiles, deciles, percentiles, etc., as special cases, is *quantile*.

Chapter 3
Collecting good data

Raw data, like raw potatoes, usually require cleaning before use.

Ronald A. Thisted

Data provide a window to the world, but it is important that they give us a clear view. A window with scratches, distortions, or with marks on the glass is likely to mislead us about what lies beyond, and it is the same with data. If data are distorted or corrupted in some way then mistaken conclusions can easily arise. In general, not all data are of high quality. Indeed, I might go further than this and suggest that it is rare to meet a data set which does not have quality problems of some kind, perhaps to the extent that if you encounter such a 'perfect' data set you should be suspicious. Perhaps you should ask what preprocessing the data set has been subjected to which makes it look so perfect. We will return to the question of preprocessing later.

Standard textbook descriptions of statistical ideas and methods tend to assume that the data have no problems (statisticians say the data are 'clean', as opposed to 'dirty' or 'messy'). This is understandable, since the aim in such books is to describe the methods, and it detracts from the clarity of the description to say what to do if the data are not what they should be. However, this book is rather different. The aim here is not to teach the mechanics of statistical methods, but rather to introduce and convey the flavour of the real discipline. And the real discipline of statistics has to cope with dirty data.

In order to develop our discussion, we need to understand what could be meant by 'bad data', how to recognize them, and what to do about them. Unfortunately, data are like people: they can 'go bad' in an unlimited number of different ways. However, many of those ways can be classified as either *incomplete* or *incorrect*.

Incomplete data

A data set is incomplete if some of the observations are missing. Data may be randomly missing, for reasons entirely unrelated to the study. For example, perhaps a chemist dropped a test tube, or a patient in a clinical trial of a skin cream missed an appointment because of a delayed plane, or someone moved house and so could not be contacted for a follow-up questionnaire. But the fact that a data item is missing can also in itself be informative. For example, people completing an application form or questionnaire may wish to conceal something, and, rather than lie outright, may simply not answer that question. Or perhaps only people with a particular view bother to complete a questionnaire. For example, if customers are asked to complete forms evaluating the service they have received, those with axes to grind may be more inclined to complete them. If this is not recognized in the analysis, a distorted view of customers' opinions will result. Internet surveys are especially vulnerable to this kind of thing, with people often simply being invited to respond. There is no control over how representative the respondents are of the overall population, or even if the same people respond multiple times.

Other examples of this sort of 'selection bias' abound, and can be quite subtle. For example, it is not uncommon for patients to drop out of clinical trials of medicines. Suppose that patients who recovered while using the medicine failed to return for their next appointment, because they felt it was unnecessary (since they had recovered). Then we could easily draw the conclusion that the medicine did not work, since we would see only patients who were still sick.

A classic case of this sort of bias arose when the *Literary Digest* incorrectly predicted that Landon would overwhelmingly defeat Roosevelt in the 1936 US presidential election. Unfortunately, the questionnaires were mailed only to people who had both telephones and cars, and in 1936 these people were wealthier on average than the overall population. The people sent questionnaires were not properly representative of the overall population. As it turned out, the bulk of the others supported Roosevelt.

Another, rather different kind of case of incorrect conclusions arising from failure to take account of missing data has become a minor statistical classic. This is the case of the *Challenger* space shuttle, which blew up on launch in 1986, killing everyone on board. The night before the launch, a meeting was held to discuss whether to go ahead, since the forecast temperature for the launch date was exceptionally low. Data were produced showing that there was apparently no relationship between air temperature and damage to certain seals on the booster rockets. However, the data were incomplete, and did not include all those launches involving *no* damage. This was unfortunate because the launches when no damage occurred were predominantly made at higher temperatures. A plot of *all* of the data shows a clear relationship, with damage being more likely at lower temperatures.

As a final example, people applying for bank loans, credit cards, and so on, have a 'credit score' calculated, which is essentially an estimate of the probability that they will fail to repay. These estimates are derived from statistical models built (as described in Chapter 6) using data from previous customers who have already repaid or failed to repay. But there is a problem. Previous customers are not representative of all people who applied for a loan. After all, previous customers were chosen because they were thought to be good risks. Those applicants thought to be intrinsically poor risks and likely to default would not have been accepted in the first place, and would therefore not be included in the data. Any statistical model which fails to take account of this distortion of the data set is likely to lead to mistaken conclusions. In this case, it could well mean the bank collapsing.

If only some values are missing for each record (e.g. some of the answers to a questionnaire), then there are two common elementary approaches to analysis. One is simply to discard any incomplete records. This has two potentially serious weaknesses. The first is that it can lead to selection bias distortions of the kind discussed above. If records of a particular kind are more likely to have some values missing, then deleting these records will leave a distorted data set. The second serious weakness is that it can lead to a dramatic reduction in the size of the data set available for analysis. For example, suppose a questionnaire contains 100 questions. It is entirely possible that *no* respondent answered *every* question, so that *all* records may have something missing. This means that dropping incomplete responses would lead to dropping all of the data.

The second popular approach to handling missing values is to insert substitute values. For example, suppose age is missing from some records. Then we could replace the missing values by the average of the ages which had been recorded. Although this results in a complete(d) data set, it also has disadvantages. Essentially we would be making up data.

If there is reason to suspect that the fact that a number is missing is related to the value it would have had (for example, if older people are less likely to give their age) then more elaborate statistical techniques are needed. We need to construct a statistical model, perhaps of the kind discussed in Chapter 6, of the probability of being missing, as well as for the other relationships in the data.

It is also worth mentioning that it is necessary to allow for the fact that not all values have been recorded. It is common practice to use a special symbol to indicate that a value is missing. For example, N/A, for 'not available'. But sometimes numerical codes are used, such as 9999 for age. In this case, failure to let the computer know that 9999 represents missing values can lead to a wildly inaccurate result. Imagine the estimated average age when there are many values of 9999 included in the calculation...

In general, and perhaps this should be expected, there is no perfect solution to missing data. All methods to handle it require some kind of additional assumptions to be made. The best solution is to minimize the problem during the data collection phase.

Incorrect data

Incomplete data is one kind of data problem, but data may be *incorrect* in any number of ways and for any number of reasons. There are both high and low level reasons for such problems.

One high level reason is the difficulty of deciding on suitable (and universally agreed) definitions. Crime rate, referred to in Chapter 1, provides an example of this. Suicide rate provides another. Typically, suicide is a solitary activity, so that no one else can know for certain that it was suicide. Often a note is left, but not in all cases, and then evidence must be adduced that the death was in fact suicide. This moves us to murky ground, since it raises the question of what evidence is relevant and how much is needed. Moreover, many suicides disguise the fact that they took their own life; for example so that the family can collect on the life insurance.

In a different, but even more complicated situation, the National Patient Safety Agency in the UK is responsible for collating reports of accidents which have occurred in hospitals. The Agency then tries to classify them to identify commonalities, so that steps can be taken to prevent accidents happening in the future. The difficulty is that accidents are reported by many thousands of different people, and described in different ways. Even the same incident can be described very differently.

At a lower level, mistakes are often made in reading instruments or recording values. For example, a common tendency in reading instruments is to subconsciously round to the nearest whole number. Distributions of blood pressure measurements recorded using old-fashioned (non-electronic) sphygmomanometers show a clear tendency for more values to be recorded at 60, 70, and 80mm of mercury than at neighbouring values, such as 69 or 72. As far as recording errors go, digits may be transposed (28, instead of 82); the handwritten digit 7 may be mistaken for 1 (less likely in continental Europe, where 7 is written 7̶); data may be put in the wrong column on a form, so accidentally multiplying values by 10; the US style of date (month/day/year) might be confused with the UK style (day/month/year) or vice versa; and so on. In 1796, the Astronomer Royal Nevil Maskelyne dismissed his assistant, David Kinnebrook, on the grounds that the latter's observations of the times at which a chosen star crossed the meridian wire in a telescope at Greenwich were too inaccurate. This mattered, because the accuracy of the clock at Greenwich hinged on accurate measurements of the transit times, estimates of the longitude of the nation's ships depended on the clock, and the British Empire depended on its ships. However, later investigators have explained the inaccuracies in terms of psychological reaction time delays and the subconscious rounding phenomenon mentioned above. And, as a final example from the many I could have chosen, the 1970 US Census said there were 289 girls who had been both widowed and divorced by the age of 14. We should also note the general point that the larger the data set, the more hands involved in its compilation, and the more stages involved in its processing, the more likely it is to contain errors.

Other low level examples of data errors often arise with units of measurement, such as recording height in metres rather than feet, or weight in pounds rather than kilograms. In 1999, the *Climate Orbiter* Mars probe was lost when it failed to enter the Martian atmosphere at the correct angle because of confusion between pressure measurements based on pounds and on newtons. In another example of confusion of units, this time in a medical context, an elderly lady usually had normal blood calcium levels, in the range 8.6 to 9.1, which suddenly appeared to drop to a much lower value of 4.8. The nurse in charge was about to begin infusing calcium, when Dr Salvatore Benvenga discovered that the apparent drop was simply because the laboratory had changed the units in which it reported its results (from milligrams per decilitre to milliequivalents per litre).

Error propagation

Once made, errors can propagate with serious consequences. For example, budget shortfalls and possible job layoffs in Northwest Indiana in 2006 were attributed to the effect of a mistake in just one number working its way up through the system. A house that should have been valued at $121,900 had its value accidentally changed to $400 million. Unfortunately, this mistaken value was used in calculating tax rates.

In another case, the *Times* of 2 December 2004 reported how 66,500 of around 170,000 firms were accidentally removed from a list used to compile official estimates of construction output in the UK. This led to a reported fall of 2.6% in construction growth in the first quarter, rather than the correct value of an increase of 0.5%, followed by a reported growth of 5.3% rather than the correct 2.1% in the second quarter.

Preprocessing

As must be obvious from the examples above, an essential initial component of any statistical analysis is a close examination of the data, checking for errors, and correcting them if possible. In some contexts, this initial stage can take longer than the later analysis stages.

A key concept in data cleaning is that of an *outlier*. An outlier is a value that is very different from the others, or from what is expected. It is way out in the tail of a distribution. Sometimes such extreme values occur by chance. For example, although most weather is fairly mild, we do get occasional severe storms. But in other instances anomalies arise because of the sorts of errors illustrated above, such as the anemometer which apparently reported a sudden huge gust of wind every midnight, coincidentally at the same time that it automatically reset its calibration. So one good general strategy for detecting errors in data is to look for outliers, which can then be checked by a human. These might be outliers on single variables (e.g. the man with a reported age of 210), or on multiple variables, neither of which is anomalous in itself (e.g. the 5-year-old girl with 3 children).

Of course, outlier detection is not a universal solution to detecting data errors. After all, errors can be made that lead to values which appear perfectly normal. Someone's sex may mistakenly be coded as male instead of female. The best answer is to adopt data-entry practices that minimize the number of errors. I say a little more about this below.

If an apparent error is detected, there is then the problem of what to do about it. We could drop the value, regarding it as missing, and then try to use one of the missing value procedures mentioned above. Sometimes we can make an intelligent guess as to what the value should have been. For example, suppose that, in recording the ages of a group of students, one had obtained the string of values 18, 19, 17, 21, 23, 19, 210, 18, 18, 23. Studying these, we might think it likely that the 210 had been entered into a wrong column, and that it should be 21. By the way, note the phrase 'intelligent guess' used above. As with all statistical data analysis, careful thought is crucial. It is not simply a question of choosing a particular statistical method and letting the computer do the work. The computer only does the arithmetic.

The example of student ages in the previous paragraph was very small, just involving ten numbers, so it was easy to look through them, identify the outlier, and make an intelligent guess about what it should have been. But we are increasingly faced with larger and larger data sets. Data sets of many billions of values are commonplace nowadays in scientific applications (e.g. particle experiments), commercial applications (e.g. telecommunications), and other areas. It will often be quite infeasible to explore all the values manually. We have to rely on the computer. Statisticians have developed automatic procedures for detecting outliers, but these do not completely solve the problem. Automatic procedures may raise flags about certain kinds of strange values, but they will ignore peculiarities they have not been told about. And then there is the question of what to do about an apparent anomaly detected by the computer. This is fine if only 1 in those billion numbers is flagged as suspicious, but what if 100,000 are so flagged? Again, human examination and correction is impracticable. To cope with such situations, statisticians have again developed automated procedures. Some of the earliest such automated editing and correcting methods were developed in the context of censuses and large surveys. But they are not foolproof. The bottom line is, I am afraid, once again, that statisticians cannot work miracles. Poor data risk yielding poor (meaning inaccurate, mistaken, error-prone) results. The best strategy for avoiding this is to ensure good-quality data from the start.

Many strategies have been developed for avoiding errors in data in the first place. They vary according to the application domain and the mode of data capture. For example, when clinical trial data are copied from hand-completed case record forms, there is a danger of introducing errors in the transcription phase. This is reduced by arranging for the exercise to be repeated twice, by different people working independently, and then checking any differences. When applying for a loan, the application data (e.g. age, income, other debts, and so on) may be entered directly into a computer, and interactive computer software can cross-check the answers as they are given (e.g. if a house owner, do the debts include a mortgage?). In general, forms should be designed so as to minimize errors. They should not be excessively complicated, and all questions should be unambiguous. It is obviously a good idea to conduct a small pilot survey to pick up any problems with the data capture exercise before going live.

Incidentally, the phrase 'computer error' is a familiar one, and the computer is a popular scapegoat when data mistakes are made. But the computer is just doing what it is told, using the data provided. When errors are made, it is not the computer's fault.

Observational versus experimental data

It is often useful to distinguish between *observational* and *experimental* studies, and similarly between observational and experimental data. The word 'observational' refers to situations in which one cannot interfere or intervene in the process of capturing the data. Thus, for example, in a survey (see below) of people's attitudes to politicians, an appropriate sample of people would be asked how they felt. Or, in a study of the properties of distant galaxies, those properties would be observed and recorded. In both of these examples, the researchers simply chose who or what to study and then recorded the properties of those people or objects. There is no notion of doing something to the people or galaxies before measuring them. In contrast, in an experimental study the researchers would actually manipulate the objects in some way. For example, in a clinical trial they might expose volunteers to a particular medication, before taking the measurements. In a manufacturing experiment to find the conditions which yield the strongest finished product, they would try different conditions.

One fundamental difference between observational and experimental studies is that experimental studies are much more effective at sorting out what causes what. For example, we might conjecture that a particular way of teaching children to read (method A, say) is much more effective than another (method B). In an observational study, we will look at children who have been taught by each method, and compare their reading ability. But we will not be able to influence who is taught by method A and who by method B; this is determined by someone else. This raises a potential problem. It means that it is possible that there are other differences between the two reading groups, as well as teaching method. For example, to take an extreme illustration, a teacher may have assigned all the faster learners to method A. Or perhaps the children themselves were allowed to choose, and those already more advanced in reading tended to choose method A. If we are a little more sophisticated in statistics, we might use statistical methods to try to control for any pre-existing differences between the children, as well as other factors we think are likely to influence how quickly they would learn to read. But there will always remain the possibility that there are other influences we have not thought of which cause the difference.

Experimental studies overcome this possibility by deliberately choosing which child is taught by each method. If we did know all the possible factors, in addition to teaching method, which could influence reading ability, we could make sure that the assignment to teaching method was 'balanced'. For example, if we thought that reading ability was influenced by age, we could assign the same number of young children to each method. By this means, any differences in reading ability arising from age would have no impact on the difference between our two groups: if age did influence reading ability, the impact would be the same in each group. However, as it happens, experimental studies have an even more powerful way of choosing which child receives which method, called *randomization*. I discuss this below.

The upshot of this is that, in an experimental study we can be more confident of the cause of any observed effect. In the experiment comparing teaching reading, we can be more confident that any difference between the reading ability in the two groups is a consequence of the teaching method, rather than of some other factor.

Unfortunately it is not always possible to conduct experiments rather than observational studies. We do not have much opportunity to expose different galaxies to different treatments! In any case, sometimes it would be misleading to use an experimental approach: in many social surveys, the aim is to find out what the population is actually like, not 'what would be the effect if we did such and such'. However, if we do want to know what would be the effect of a potential intervention, then experimental studies are the better strategy. They are universal in the pharmaceutical sector, very widespread in medicine and psychology, ubiquitous in industry and manufacturing, and increasingly used to evaluate social policy and in areas such as customer value management.

In general, when collecting data with the aim of answering or exploring certain questions, the more data that are collected, the more accurate an answer that can be obtained. This is a consequence of *the law of large numbers*, discussed in Chapter 4. But collecting more data incurs greater cost. It is therefore necessary to strike a suitable compromise between the amount of data collected and the cost of collecting it. Various subdisciplines of statistics are central to this exercise. In particular, *experimental design* and *survey sampling* are two key disciplines.

Experimental design

We have already seen examples of very simple experiments. One of the simplest is a two-group randomized clinical trial. Here the aim is to compare two alternative treatments (A and B, say) so that we can say which of the two should be given to a new patient. To explore this, we give treatment A to one sample of patients, treatment B to another sample of patients, and evaluate the treatments' effectiveness. If, on average, A beats B, then we will recommend that the new patient receives treatment A. The meaning of the word 'beats' in the previous sentence will depend on the precise study. It could mean 'cures more patients', 'extends average lifespan', 'yields greater average reduction in pain', and so on.

Now, as we have already noted above, if the two groups of patients differ in some way, then the conclusions we can draw are limited. If those who received treatment A were all male, and those who received treatment B were all female, then we would not know if any difference between the groups that we observed was due to the treatment or to the sex difference: maybe females get better faster, regardless of treatment. The same point applies to any other factor – age, height, weight, duration of illness, previous treatment history, and so on.

One strategy to alleviate this difficulty is to randomly allocate patients to the two treatment groups. The strength of this approach is that, while it does not guarantee balance (e.g., it is possible that this random allocation procedure might lead to a substantially higher proportion of males in one group than the other), basic rules of probability (discussed in Chapter 4) tell us that large imbalances are extremely unlikely. In fact, it is possible to go further than this and work out just how likely different degrees of imbalance are. This in turn enables us to calculate how confident we should be in our conclusions.

Moreover, if the random allocation is *double blind*, there is no risk of subconscious bias creeping into the allocation or the measurement of patients. A study is double blind if neither the patient nor the doctor conducting the trial knows which treatment the patient is receiving. This can be achieved by making the tablets or medicines look identical, and simply coding them as X or Y without indicating which of the treatments is which. Only later, after the analysis has revealed that X is better than Y, is the coding broken, to show that X is really treatment A or B as the case may be.

The two-group randomized clinical trial is very simple, and has obvious extensions: for example, we can immediately extend it to more than two treatment groups. However, for the sake of variety, I shall switch examples. A market gardener might want to know which of low and high levels of water is better, in terms of producing greater crop yield. He could conduct a simple two-group experiment, of the kind described above, to determine this. Since we know that outcomes are not totally predictable, he will want to expose more than one greenhouse to the low level of water, and more than one to the high level, and then calculate the average yields at each level. He might, for example, decide to use four greenhouses for each level. This is precisely the same sort of design as in the teaching methods study above.

But now suppose that the farmer also wants to know which of low and high levels of fertilizer is more effective. The obvious thing to do is to conduct another two-group experiment, this time with four greenhouses receiving the low level of fertilizer and four receiving the high level. This is all very well, but to answer both of the questions, the water one and the fertilizer one, requires a total of sixteen greenhouses. If the farmer is also interested in the effectiveness of low and high levels of humidity, temperature, hours of sunlight, and so on, we see that we will soon run out of greenhouses.

Now, there is a very clever way round this, using the notion of a *factorial* experimental design. Instead of carrying out two separate experiments, one for water and one for fertilizer, the farmer can treat two greenhouses with (fertilizer = low, water = low), two with (low, high), two with (high, low), and two with (high, high). This requires just eight greenhouses, and yet we are still treating four of them with the low water level and four with the high water level, as well as four with the low fertilizer level and four with the high fertilizer level, so that the results of the analysis will be just as accurate as when we did two separate experiments.

In fact, this factorial design (each of water and fertilizer is a 'factor') has an additional attractive feature. It allows us to see if the impact of the level of fertilizer is different at the two levels of water: perhaps the difference between yields with the low and high levels of fertilizer varies between the two levels of water. This so-called *interaction* effect cannot be examined in the two separate experiments approach.

This basic idea has been extended in many ways to yield very powerful tools for obtaining accurate information for the minimum cost. When combined with other experimental design tools, such as balance, randomization, and controlling for known influences, some highly sophisticated experimental designs have been developed.

Sometimes, in experiments, non-statistical issues are important. For example, in clinical trials and other medical and social policy investigations, ethical issues may be relevant. In a clinical trial comparing a proposed new treatment against an (inactive) placebo, we will know that half of the volunteer patients will receive something which has no biological impact. Is that appropriate? Is there a danger that those exposed to the proposed new treatment might suffer from side effects? Such things have to be balanced against the fact that untold numbers of future patients will benefit from what is learned in the trial.

Survey sampling

Imagine that, in order to run the country effectively, we wish to know the average income of the one million employed men and women in a certain town. In principle, we could determine this by asking each of them what their income was, and averaging the results. In practice, this would be extremely difficult, verging on the impossible. Apart from anything else, over the course of the time taken to collect the data it is likely that incomes would change: some people would have left or changed their jobs, others would have received raises, and so on. Furthermore, it would be extremely costly tracking down each person. We might try to reduce costs by relying on the telephone, rather than face to face interviews. However, as we have already seen, in the extreme case of the 1936 US presidential election, there is a great risk that we would miss important parts of the population.

What we need is some way to reduce the cost of collecting the data while at the same time making the process quicker and, if possible, also more accurate. Put this way, it probably sounds like a tall order, but statistical ideas and tools that have these properties do exist. The key idea is one we have met several times before: the notion of a sample.

Suppose that, instead of finding out what each of the one million employees earned, we simply asked a thousand of them. Now clearly we have to be careful about exactly which thousand we ask. The reasons are essentially the same as when we were designing a simple two-group experiment and had to take steps to ensure that the only difference between the groups was that one received treatment A and one received treatment B. Now we have to ensure that the particular thousand people we approach are *representative* of the full population of a million.

What do we mean by 'representative'? Ideally, our sample of a thousand should have the same proportion of men in it as the entire population, the same number of young people, the same number of part-time workers, and so on. To some extent we can ensure this, choosing the thousand so that the proportion of men is correct, for example. But there is obviously a practical limit to what we can deliberately balance in this way.

We saw how to handle this when we looked at experimental design. There we tackled the difficulty by *randomly allocating* patients to each group. Here we tackle it by *randomly sampling* the thousand people from the total population. Once again, while this does not guarantee that the sample will be similar in composition to the entire population, basic probability tells us that the chance of obtaining a seriously dissimilar sample is very small. In particular, it follows that the probability that our estimate of the average income, derived from the sample, will be very different from the average income in the entire population is very small. Indeed, two properties of probability which we will explore later, the *law of large numbers* and the *Central Limit Theorem* also tell us that we can make this probability as small as we like by increasing the sample size. It turns out that what matters is not how large a fraction of the population is included in the sample, but simply how large the sample is. Our estimate, based on a sample size of one thousand, would essentially be just as accurate if the entire population consisted of ten million or ten billion people. Since sample size is directly related to the cost of collecting the data, we now have an immediate relationship between accuracy and cost: the larger our sample the greater the cost but the smaller the probability of significant deviation between the sample estimate and the overall population average.

While 'randomly sampling a thousand people from the population' of employed people in the town may sound like a simple exercise, in fact it takes considerable care. We cannot, for example, simply choose the thousand people from the largest employer in the town, since these may not be representative of the overall million. Likewise, we cannot call at a random sample of people's homes at

8pm in the evening, since we would miss those who worked late, and these workers may differ in average income from the others. In general, to ensure that our sample of a thousand is properly representative we need a *sampling frame*, a list of all the one million employed people in our population, from which we can randomly choose a thousand. Having such a list ensures that everyone is equally likely to be included.

This notion of *simple random sampling* is the basic idea behind survey sampling. We draw up a sampling frame and from it randomly choose the people to be included in our sample. We then track them down (interview, phone, letter, email, or whatever) and record the data we want. This basic idea has been elaborated in many very sophisticated and advanced ways, yielding more accurate and cheaper approaches. For example, if we intended to interview each of the thousand respondents it could be quite costly in terms of time and travel expenses. It would be better, from this perspective, to choose respondents from small geographically local clusters. *Cluster sampling* extends simple random sampling by allowing this. Instead of randomly choosing a thousand people from the entire population, it selects (say) ten groups of a hundred people each, with the people in each group located near to each other. Likewise, we can be certain that balance is achieved on some factors, rather than simply relying on the random sampling procedure, if we enforce the balance in the way we choose the sample. For example, we could randomly choose a number of women from the population, and separately randomly choose a number of men from the population, where the numbers are chosen so that the proportions of males and females are the same as in the population. This procedure is known as *stratified sampling*, since it divides the overall population listed in the sampling frame into strata (men and women in this case). If the variable used for the stratification (sex in this example) is strongly related to the variable we are interested in (here, income), then this can yield improved accuracy for the same sample size.

In general, in survey sampling, we are very lucky if we obtain responses from everyone approached. Almost always there is some non-response. We are back to the missing data problem discussed earlier, and, as we have seen, missing data can lead to a biased sample and incorrect conclusions. If those earning large salaries refused to reply, then we would underestimate the average income in the population. Because of this, survey experts have developed a wide range of methods of minimizing and adjusting for non-response, including repeated call-backs to non-responders and statistical reweighting procedures.

Conclusion

This chapter has described the raw material of statistics, the data. Sophisticated data collection technologies have been developed by statisticians to maximize the information obtained for the minimum cost. But it would be naive to believe that perfect data can usually be obtained. Data are a reflection of the real world, and the real world is complicated. Recognizing this, statisticians have also developed tools to cope with poor-quality data. But it is important to recognize that statisticians are not magicians. The old adage of 'garbage in, garbage out' is just as true in statistics as elsewhere.

Chapter 4
Probability

Being a statistician means never having to say you are certain.

Anon

The essence of chance

One of the definitions of statistics given in Chapter 1 was that it is the science of handling uncertainty. Since it is abundantly clear that the world is full of uncertainty, this is one reason for the ubiquity of statistical ideas and methods. The future is an unknown land and we cannot be certain about what will happen. The unexpected does occur: cars break down, we have accidents, lightning does strike, and, lest I am giving the impression that such things are always bad, people do even win lotteries. More prosaically, it is uncertain which horse will win the race or which number will come up on the throw of a die. And, at the end of it all, we cannot predict exactly how long our lives will be.

However, notwithstanding all that, one of the greatest discoveries mankind has made is that there are certain principles covering chance and uncertainty. Perhaps this seems like a contradiction in terms. Uncertain events are, by their very nature, uncertain. How, then, can there be natural laws governing such things?

One answer is that while an individual event may be uncertain and unpredictable, it is often possible to say something about collections of events. A classic example is the tossing of a coin. While I cannot say whether a coin will come up heads or tails on a particular toss, I can say with considerable confidence that if I toss the coin many times then around half of those times it will show heads and around half tails. (I am assuming here that the coin is 'fair', and that no sleight of hand is being used when tossing it.) Another example in the same vein is whether a baby will be male or female. It is, on conception, a purely chance and unpredictable event which gender the child will become. But we know that over many births just over a half will be male.

This observable property of nature is an example of one of the laws governing uncertainty. It is called the *law of large numbers* because of the fact that the proportion gets closer and closer to a particular value (a half in the cases of the fair coin and of babies' gender) the more cases we consider. This law has all sorts of implications, and is one of the most powerful of statistical tools in taming, controlling, and allowing us to take advantage of uncertainty. We return to it later in this chapter, and repeatedly throughout the book.

Understanding probability

So that we can discuss matters of uncertainty and unpredictability without ambiguity, statistics, like any other scientific discipline, uses a precise language: the language of *probability*. If this is your first exposure to the language of probability, then you should be warned that, as with one's first exposure to any new language, some effort will be required to understand it. Indeed, bearing that in mind, you might find that this chapter requires more than one reading: you might like to reread this chapter once you have reached the end of the book.

Development of the language of probability blossomed in the 17th century. Mathematicians such as Blaise Pascal, Pierre de Fermat, Christiaan Huygens, Jacob Bernoulli, and later Pierre Simon Laplace, Abraham De Moivre, Siméon-Denis Poisson, Antoine Cournot, John Venn, and others laid its foundations. By the early 20th century, all the ideas for a solid science of probability were in place, and in 1933 the Russian mathematician Andrei Kolmogorov presented a set of axioms which provided a complete formal mathematical *calculus* of probability. Since then, this axiom system has been almost universally adopted.

Kolmogorov's axioms provide the machinery by which to manipulate probabilities, but they are a mathematical construction. To use this construction to make statements about the real world, it is necessary to say what the symbols in the mathematical machinery represent in that world. That is, we need to say what the mathematics 'means'.

The probability calculus assigns numbers between 0 and 1 to uncertain events to represent the probability that they will happen. A probability of 1 means that an event is certain (e.g. the probability that, if someone looked through my study window while I was writing this book, they would have seen me seated at my desk). A probability of 0 means that an event is impossible (e.g., the probability that someone will run a marathon in ten minutes). For an event that *can* happen but is neither certain nor impossible, a number between 0 and 1 represents its 'probability' of happening.

One way of looking at this number is that it represents the *degree of belief* an individual has that the event will happen. Now, different people will have more or less information relating to whether the event will happen, so different people might be expected to have different degrees of belief, that is different probabilities for the event. For this reason, this view of probability is called *subjective* or *personal* probability: it depends on who is assessing the probability. It is also clear that someone's probability might change as more information becomes available. You might start with a probability, a degree of belief, of 1/2 that a particular coin will come up heads (based on your previous experience with other tossed coins), but after observing 100 consecutive heads and no tails appear you might become suspicious and change your subjective probability that this coin will come up heads.

Tools have been developed to estimate individuals' subjective probabilities based on betting strategies, but, as with any measurement procedure, there are practical limitations on how accurately probabilities can be estimated.

A different view of the probability of an event is that it is the proportion of times the event would happen if identical circumstances were repeated an infinite number of times. The fair coin tossing example above is an illustration. We have seen that, as the coin is tossed, so the proportion of heads gets closer and closer to some specific value. This value is defined as the probability that the coin will come up heads on any single toss. Because of the role of frequencies, or counts, in defining this interpretation of probability, it is called the *frequentist* interpretation.

Just as with the subjective approach, there are practical limitations preventing us from finding the exact frequentist probability. Two tosses of a coin cannot really have *completely* identical circumstances. Some molecules will have worn from the coin in the first toss, air currents will differ, the coin will have been slightly warmed by contact with the fingers the first time. And in any case we have to stop tossing the coin sometime, so we cannot actually toss it an infinite number of times.

These two different interpretations of what is meant by probability have different properties. The subjective approach can be used to assign a probability to a unique event, something about which it makes no sense to contemplate an infinite, or even a large number of repetitions under identical circumstances. For example, it is difficult to know what to make of the suggestion of an infinite sequence of identical attempts to assassinate the next president of the USA, with some having one outcome and some another. So it seems difficult to apply the frequentist interpretation to such an event. On the other hand, the subjective approach shifts probability from being an objective property of the external world (like mass or length) to being a property of the interaction between the observer and the world. Subjective probability is, like beauty, in the eye of the beholder. Some might feel that this is a weakness: it means that different people could draw different conclusions from the same analysis of the same data. Others might regard it as a strength: the conclusions would have been influenced by your prior knowledge.

There are yet other interpretations of probability. The 'classical' approach, for example, assumes that all events are composed of a collection of equally likely elementary events. For example, a throw of a die might produce a 1, 2, 3, 4, 5, or 6 and the symmetry of the die suggests these six outcomes are equally likely, so each has a probability of 1/6 (they must sum to 1, since it is *certain* that one of 1, 2, 3, 4, 5, or 6 will come up). Then, for example, the probability of getting an even number is the sum of the probabilities of each of the equally likely events of getting a 2, a 4, or a 6, and is therefore equal to 1/2. In less artificial circumstances, however, there are difficulties in deciding what these 'equally likely' events are. For example, if I want to know the probability that my morning journey to work will take less than one hour, it is not at all clear what the equally likely elementary events should be. There is no obvious symmetry in the situation, analogous to that of the die. Moreover, there is the problem of the circular content of the definition in requiring the elementary events to be 'equally likely'. We seem to be defining probability in terms of probability.

It is worth emphasizing here that all of these different interpretations of probability conform to the same axioms and are manipulated by the same mathematical machinery. It is simply the mapping to the real world which differs; the definition of what the mathematical object *means*. I sometimes say that the *calculus* is the same, but the *theory* is different. In statistical applications, as we will see in Chapter 5, the different interpretations can sometimes lead to different conclusions being drawn.

The laws of chance

We have already noted one law of probability, the law of large numbers. This is a law linking the mathematics of probability to empirical observations in the real world. Other laws of probability are implicit in the axioms of probability. Some very important laws involve the concept of *independence*.

Two events are said to be independent if the occurrence of one does not affect the probability that the other will occur. The fact that a coin tossed with my left hand comes up tails rather than heads does not influence the outcome of a coin tossed with my right hand. These two coin tosses are independent. If the probability is 1/2 that the coin in my left hand will come up heads, and the probability is 1/2 that the coin in my right hand will come up heads, then the probability that both will come up heads is $1/2 \times 1/2 = 1/4$. This is easy to see since we would expect that in many repetitions of the double tossing experiment we would obtain about half of the left hand coins showing heads, and, *amongst those*, about half of the right hand coins would show heads because the outcome of the first toss does not influence the second. Overall, then, about 1/4 of the double tosses would show two heads. Similarly, about 1/4 would show left tails, right heads, about 1/4 would show left heads, right tails, and about 1/4 would show both left and right tails.

In contrast, the probability of falling over in the street is certainly not independent of whether it has snowed; these events are *dependent*. We saw another example of dependent events in Chapter 1: the tragic Sally Clark case of two cot deaths in the same family. When events are not independent, we cannot calculate the probability that both will happen simply by multiplying together their separate probabilities. Indeed, this was the mistake which lay at the root of the Sally Clark case. To see this, let us take the most extreme situation of events which are completely dependent: that is, when the outcome of one *completely determines* the outcome of the other. For example, consider a single toss of a coin, and the two events 'the coin faces heads up' and 'the coin faces tails down'. Each of these events has a probability of a half: the probability that the coin will show heads up is 1/2, and the probability that the coin will show tails down is 1/2. But they are clearly not independent events. In fact, they are completely dependent. After all, if the first event is true (heads up) the second *must be* true (tails down). Because they are completely dependent, the probability that they will both occur is simply the probability that the first will occur – a probability of a half. This is not what we get if we multiply the two separate probabilities of a half together.

In general, dependence between two events means that the probability that one will occur depends on whether or not the other has occurred.

Statisticians call the probability that two events will *both* occur the *joint probability* of those two events. For example, we can speak of the joint probability that I will slip over *and* that it snowed. The joint probability of two events is closely related to the probability that an event will occur *if* another one has occurred. This is called the *conditional probability* – the probability that one event will occur given that we know that the other one has occurred. Thus we can talk of the conditional probability that I will slip over, *given that* it snowed.

The (joint) probability that both events A and B occur is simply the probability that A occurs times the (conditional) probability that B occurs given that A occurs. The (joint) probability that it snows and I slip over is the probability that it snows times the (conditional) probability that I slip over if it has snowed.

To illustrate, consider a single throw of a die, and two events. Event A is that the number showing is divisible by 2, and Event B is that the number showing is divisible by 3. The joint probability of these two events A and B is the probability that I get a number which is both divisible by 2 and is divisible by 3. This is just 1/6, since only one of the numbers 1, 2, 3, 4, 5, and 6 is divisible by both 2 and 3. Now, the conditional probability of B given A is the probability that I get a number which is divisible by 3 *amongst those that are divisible by 2*. Well, amongst all the numbers which are divisible by 2 (that is, amongst 2, 4, or 6) only one is divisible by 3, so this conditional probability is 1/3. Finally, the probability of event A is 1/2 (half of the numbers 1, 2, 3, 4, 5, and 6 are divisible by 2). We therefore find that the probability of A (1/2) times the (conditional) probability of B given A (1/3) is 1/6. This is the same as the joint probability of obtaining a number divisible by both 2 and 3; that is, the joint probability of events A and B both occurring.

In fact, we previously met the concept of conditional probability in Chapter 1, in the form of the Prosecutor's Fallacy. This pointed out that the probability of event A occurring given that event B had occurred was not the same as the probability of event B occurring given that event A had occurred. For example, the probability that someone who runs a major corporation can drive a car is not the same as the probability that someone who can drive a car runs a major corporation. This leads us to another very important law of probability: *Bayes's theorem* (or *Bayes's rule*). Bayes's theorem allows us to relate these two conditional probabilities, the conditional probability of A given B and the conditional probability of B given A.

We have just seen that the probability that both events A and B will occur is equal to the probability that A will occur, times the (conditional) probability that B will occur given that A has occurred. But this can also be written the other way round: the probability that both events A and B will occur is also equal to the probability that B will occur times the probability that A will occur given that B has occurred. All Bayes's theorem says (though it is usually expressed in a different way) is that these are simply two alternative ways of writing the joint probability of A and B. That is, the probability of A times the probability of B given A is equal to the probability of B times the probability of A given B. Both are equal to the joint probability of A and B. In our 'car-driving corporate head' example, Bayes's theorem is equivalent to saying that the probability of running a major corporation given that you can drive a car, times the probability that you can drive a car, is equal to the probability that you can drive a car given that you are a corporate head, times the probability of being a corporate head. Both equal the joint probability of being a corporate head *and* being able to drive a car.

Another law of probability says that if either one of two events can occur, but not both together, then the probability that one *or* the other will occur is the sum of the separate probabilities that each will occur. If I toss a coin, which obviously cannot show heads and tails simultaneously, then the probability that a head *or* tail will show is the sum of the probability that a head will show and the probability that a tail will show. If the coin is fair, each of these separate probabilities is a half, so that the overall probability of a head or a tail is 1. This makes sense: 1 corresponds to certainty and it is certain that a head or a tail must show (I am assuming the coin cannot end up on its edge!). Returning to our die-throwing example: the probability of getting an even number was the sum of the probabilities of getting one of 2, or 4, or 6, because none of these can occur together (and there are no other ways of getting an even number on a single throw of the die).

Random variables and their distributions

We saw, in Chapter 2, how simple summary statistics may be used to extract information from a large collcction of values of some variable, condensing the collection down so that a distribution of values could be easily understood. Now, any real data set is limited in length – it can contain only a finite number of values. This finite set might be the values of *all* objects of the type we are considering (e.g. the scores of all major league football players in a certain year) or it might be the values of just some, a *sample*, of the objects. We saw examples of this when we looked at survey sampling.

A sample is a subset of the complete 'population' of values. In some cases, the complete population is ill-defined, and possibly huge or even infinite, so we have no choice but to work with a sample. For example, in experiments to measure the speed of light, each time I take a measurement I expect to get a slightly different valuc, simply due to the inaccuracies of the measurement process. And I could, at least in principle, go on taking measurements for ever; that is, the potential population of measurements is infinite. Since this is impossible, I must be content with a finite sample of measurements. Each of these measurements will be drawn from the population of values I could possibly have obtained. In other cases, the complete population is finite. For example, in a study of obesity amongst males in a certain town, the population is finite and, while in principle I might be able to weigh every man in the town, in practice I would probably not want to, and would work with a sample. Once again, each value in my sample is drawn from the population of possible values.

In both of these examples, all I know before I take each measurement is that it will have some value from the population of possible values. Each value will occur with some probability, but I cannot pin it down more than that, and I may not know what that probability is. I certainly cannot say exactly what value I will get in the next speed of light measurement or what will be the weight of the next man I measure. Similarly, in a throw of a die, I know that the outcome can be 1, 2, 3, 4, 5, or 6, and here I know that these are equally likely (my die is a perfect cube), but beyond that I cannot say which will come up. Like the speed and weight measurements, the outcome is random. For this reason such variables are called *random variables.*

We have already met the concept of quantiles. For example, in the case of percentiles, the 20*th* percentile of a distribution is the value such that 20% of the data values are smaller, the 8*th* percentile the value such that 8% of the data values are smaller, and so on. In general, the kth percentile has k% of the sample values smaller than it. And we can imagine similar percentiles defined, not merely for the sample we have observed, but for the complete population of values we could have observed. If we knew the 20*th* percentile for the complete population of values, then we would know that a value randomly taken from that population had a probability of 0.20 of being smaller than this percentile. In general, if we knew *all* the percentiles of a population, we would know the probability of drawing a value in the bottom 10%, or 25%, or 16%, or 98%, or any other percentage we cared to choose. In a sense, then, we would know everything about the distribution of possible values which we could draw. We would not know what value would be drawn next, but we would know the probability that it would be in the smallest 1% of the values in the population, in the smallest 2%, and so on.

There is a name for the complete set of quantiles of a distribution. It is called the *cumulative probability distribution.* It is a 'probability distribution' because it tells us the *probability* of drawing a value lower than any value we care to choose. And it is 'cumulative' because, obviously, the probability of drawing a value less than some value x gets larger the larger x is. In the example of the weights of males, if I know that the probability of choosing a man weighing less than 70kg is 1/2, then I know that the probability of choosing a man weighing less than 80kg is more than 1/2 because I can choose from all those weighing less than 70kg as well as those weighing between 70kg and 80kg. At the limit, the probability of drawing a value less than or equal to the largest value in the population is 1; it is a certain event.

This idea is illustrated in Figure 2. In this figure, the values of the random variable (think of weight) are plotted on the horizontal axis, and the probability of drawing smaller values is plotted on the vertical axis. The curve shows, for any given value of the random variable, the probability that a randomly chosen value will be smaller than this given value.

The cumulative probability distribution of a random variable tells us the probability that a randomly chosen value will be *less* than any given value. An alternative way to look at things is to look at the probability that a randomly chosen value will lie *between* any two given values. Such probabilities are conveniently represented in terms of areas between two values under a curve of the *density*

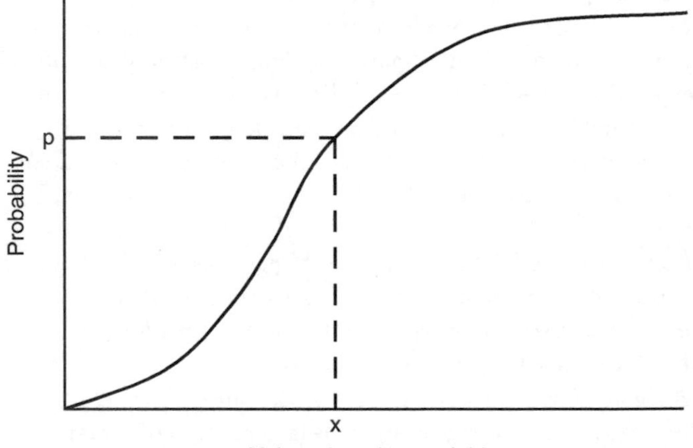

2. A cumulative probability distribution

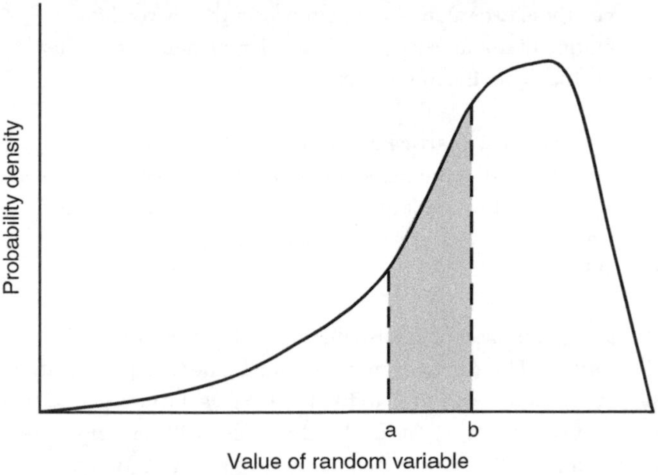

3. A probability density function

of the probability. For example, Figure 3, shows such a *probability density* curve, with the (shaded) area under the curve between points *a* and *b* giving the probability that a randomly chosen value will fall between *a* and *b*. Using such a curve for the distribution of weights of men in our town, for example, we could find the probability that a randomly chosen man would lie between 70kg and 80kg, or any other pair of values, or above or below any value we wanted. In general, randomly chosen values are more likely to occur in regions where the probability is most dense; that is, where the probability density curve is highest.

Note that the total area under the curve in Figure 3 must be 1, corresponding to certainty: a randomly chosen value must have *some* value.

Distribution curves for random variables have various shapes. The probability that a randomly chosen woman will have a weight between 70kg and 80kg will typically not be the same as the probability that a randomly chosen man will have a weight between these two values. We might expect the curve of the distribution of women's weights to take larger values at smaller weights than does the men's curve.

Certain shapes have particular importance. There are various reasons for this. In some cases, the particular shapes, or very close approximations to them, arise in natural phenomena. In other cases, the distributions arise as consequences of the laws of probability.

Perhaps the simplest of all distributions is the *Bernoulli distribution*. This can take only two values, one with probability p, say, and the other with probability $1 - p$. Since it can take only two values, it is *certain* that one or the other value will come up, so the probabilities of these two outcomes have to sum to 1. We have already seen examples illustrating why this distribution is useful: situations with only two outcomes are very common – the coin toss, with outcomes head or tail, and births, with outcomes male or female. In these two cases, p had the value 1/2 or nearly 1/2. But a huge number of other situations arise in which there are only two possible outcomes: yes/no, good/bad, default or not, break or not, stop/go, and so on.

The *binomial distribution* extends the Bernoulli distribution. If we toss a coin three times, then we may obtain no, one, two, or three heads. If we have three operators in a call centre, responding independently to calls as they come in, then none, one, two, or all three may be busy at any particular moment. The binomial distribution tells us the probability that we will obtain each of those numbers, 0, 1, 2, or 3. Of course, it applies more generally, not just to the total from three events. If we toss a coin 100 times, then the binomial distribution also tells us the probabilities that we will obtain each of 0, 1, 2, . . . , 100 heads.

Emails arrive at my computer at random. On average, during a working morning, about (say) five an hour arrive, but the number arriving in each hour can deviate from this very substantially: sometimes ten arrive, occasionally none do. The *Poisson distribution* can be used to describe the probability distribution of the number of emails arriving in each hour. It can tell us the probability (if emails arrive independently and the overall rate at which they arrive is constant) that none will arrive, that one will, that two will, and so on. This differs from the binomial distribution because, at least in principle, there is no upper limit on the number which could arrive in any hour. With the 100 coin tosses, we could not observe more than 100 heads, but I could (on a very bad day!) receive more than 100 emails in one hour.

So far, all the probability distributions I have described are for *discrete* random variables. That is, the random variables can take only certain values (two values in the Bernoulli case, counts up to the number of coin tosses/operators in the binomial case, the integers 0, 1, 2, 3, ... in the Poisson case). Other random variables are *continuous*, and can take any value from some range. Height, for example, can (subject to the accuracy of the measuring instrument) take any value within a certain range, and is not restricted to, for example, 4′, 5′, or 6′.

If a random variable can take values only within some finite interval (e.g. between 0 and 1) and if it is *equally likely* that it will take any of the values in that interval, then it is said to follow a *uniform distribution*. For example, if the postman always arrives between 10am and 11am, but in a totally unpredictable way (he is as likely to arrive between 10:05 and 10:10 as in any other five minute interval, for example), the distribution of his arrival time within this interval would be uniform.

Some random variables can take any positive value; perhaps, for example, the time duration of some phenomenon. As an illustration, consider how long glass vases survive before getting broken. Glass vases do not age, so it is no more likely that a particular favourite vase will be broken in the next year, if it is

80 years old, than that it will be broken in the next year, if it is only 10 years old (all other things being equal). Contrast this with the probability that an 80-year-old human will die next year compared with the probability that a 10-year-old human will die next year. For a glass vase, if it has not been smashed by time t, then the probability that it will be smashed in the next instant is the same, whatever the value of t (again, all other things being equal). Lifetimes of glass vases are said to follow an *exponential* distribution. In fact, there are huge numbers of applications of exponential distributions, not merely to the lifetimes of glass vases!

Perhaps the most famous of continuous distributions is the *normal* or *Gaussian distribution*. It is often loosely described in terms of its general shape: 'bell-shaped', as shown in Figure 4.

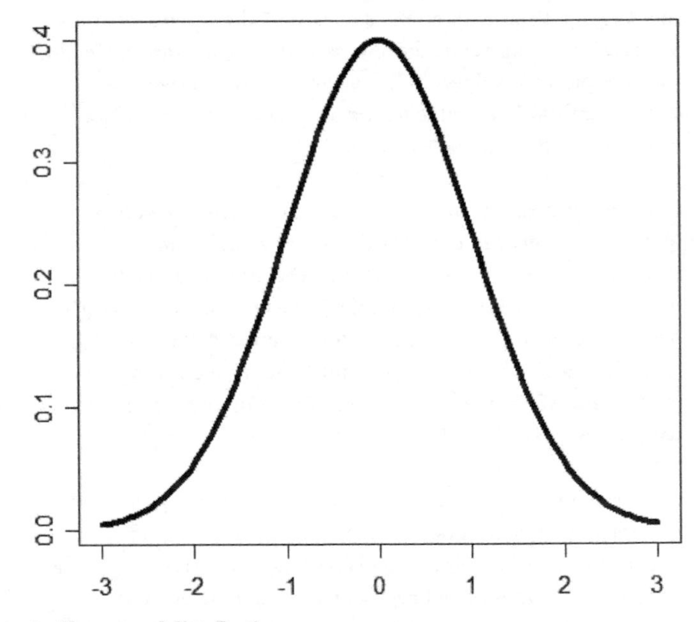

4. The normal distribution

That means that values in the middle are much more likely to occur than are values in the tails, far from the middle. The normal distribution provides a good approximation to many naturally occurring distributions. For example, the distribution of the heights of a random sample of adult men follows a roughly normal distribution.

The normal distribution also often crops up as a good model for the shape of the distribution of sample statistics (like the summary statistics described in Chapter 2) when large samples are involved. For example, suppose we repeatedly took random samples from some distribution, and calculated the means of each of these samples. Since each sample is different, we would expect each mean to be different. That is, we would have a distribution of means. If each sample is large enough, it turns out that this distribution of the means is roughly normal.

In Chapter 2, I made the point that statistics was not simply a collection of isolated tools, but was a connected language. A similar point applies to probability distributions. Although I have introduced them individually above, the fact is that the Bernoulli distribution can be seen as a special case of the binomial distribution (it is the binomial distribution when there are only two possible outcomes). Likewise, although the mathematics showing this is beyond this book, the Poisson distribution is an extreme case of the binomial distribution, the Poisson distribution and exponential distribution form a natural pair, the binomial distribution becomes more and more similar to the normal distribution the larger the maximum number of events, and so on. They are really all part of an integrated mathematical whole.

I have described the distributions above by saying that they have different shapes. In fact, these shapes can be conveniently described. We saw that the Bernoulli distribution was characterized by a value p. This told us the probability that we would get a certain outcome. Different values of p correspond to different Bernoulli distributions. We might model the outcome of a coin toss by a Bernoulli distribution with probability of heads, p, equal to a half, and model the probability of a car crash on a single journey by a Bernoulli distribution with p equal to some very small value (I hope!). In such a situation, p is called a *parameter*.

Other distributions are also characterized by parameters, serving the same role of telling us exactly which member of a family of distributions we are talking about. To see how, let us take a step back and recall the law of large numbers. This says that if we make repeated independent observations of an event which has outcome A with probability p and outcome B with probability $1 - p$, then we should expect the proportion of times outcome A is observed to get closer and closer to p the more observations we make. This property generalizes in important ways. In particular, suppose that, instead of observing an event which had only two possible outcomes, we observed an event which could take any value from a distribution on a range of values; perhaps any value in the interval $[0,1]$, for example. Suppose that we repeatedly took sets of n measurements from such a distribution. Then the law of large numbers also tells us that we should expect the mean of the n measurements to get closer to some fixed value, the larger n is. Indeed, we can picture increasing n without limit, and in that case it makes sense to talk about the mean of an unlimited sample drawn from the distribution – and even the mean of the distribution itself. For example, using this idea we can talk about not simply the mean of 'a sample drawn from an exponential distribution', but the mean of the exponential distribution itself. And, just as different Bernoulli distributions will have different parameters p, so different exponential distributions will have different means. The mean, then, is a parameter for the exponential distribution.

In an earlier example, we saw that the exponential distribution was a reasonable model for the 'lifetimes' of glass vases (under certain circumstances). Now we can imagine that we have two populations of such vases: one consisting of solid vases made of very thick glass, and the other consisting of delicate vases made of wafer-thin glass. Clearly, on average, glasses from the former population are likely to survive longer than those from the latter population. The two populations have different parameters.

We can define parameters for other distributions in a similar way: we imagine calculating the summary statistics for samples of infinite size drawn from the distributions. For example, we could imagine calculating the means of infinitely large samples drawn from members of the normal family of distributions. Things are a little more complicated here, however, because the members of this family of distributions are not uniquely identified by a single parameter. They require two parameters. In fact, the mean and standard deviation of the distributions will do. Together they serve to uniquely identify which member of the family we are talking about.

The law of large numbers has been refined even further. Imagine drawing many sets of values from some distribution, each set being of size n. For each set calculate its mean. Then the calculated means themselves are a sample from a distribution – the distribution of possible values for the mean of a sample of size n. The *Central Limit Theorem* then tells us that the distribution of these means itself approximately follows a normal distribution, and that the approximation gets better and better the larger the value of n. In fact, more than this, it also tells us that the mean of this distribution of means is identical to the mean of the overall population of values, and that the variance of the distribution of means is only $1/n$ times the size of the variance of the distribution of the overall population. This turns out to be extremely useful in statistics, because it implies that we can estimate a population mean as accurately as we like, just by taking a large enough sample (taking n large enough), with the Central Limit Theorem telling us how large a sample we must take to achieve a high probability of being that accurate. More generally, the principle that we can get better and better estimates by taking larger samples is an immensely powerful one. We already saw one way that this idea is used in practice when we looked at survey sampling in Chapter 3.

Here is another example. In astronomy, distant objects are very faint, and observations are complicated by random fluctuations in the signals. However, if we take many pictures of the same object and superimpose them, it is as if we are averaging many measurements of the same thing, each measurement drawn from the same distribution but with some extra random component. The laws of probability outlined above mean that the randomness is averaged away, leaving a clear view of the underlying signal – the astronomical object.

Chapter 5
Estimation and inference

Statistics is applied philosophy of science.

A. P. Dawid

In Chapter 1, we saw that statistics served the dual roles of summarizing data and making inferences from data. We explored some simple tools for summarizing data in Chapter 2. In this chapter, using the concepts of probability covered in Chapter 4, we look at estimation and inference. That is, we look at methods for determining the value of quantities we cannot actually observe, and making statements about them. Here are some examples.

Example 1: To determine the speed of light, we will carry out some measurement procedure. Now, no measurement procedure is perfect, and if we were to repeat the exercise we would probably obtain a slightly different value. Repeating the measurement 100 times is likely to give us 100 slightly different values. Our aim, then, is to use this sample of values to estimate the true speed of light, untarnished by measurement error.

Example 2: In a simple randomized clinical trial, we might give a new drug to one sample of patients and a standard drug to another sample. Based on observations of the effects in these two patient groups we will want to make some statement, or inference, about the relative effectiveness of the new drug. Put another way, we would want to estimate how large a difference in the effectiveness of the two drugs we might expect if we prescribed each of the drugs to the general population of patients. We would also ideally like some indication of how confident we were in the size of the estimate.

Example 3: In studying unemployment in London, it would be infeasible to interview everyone, so a sample of people would be interviewed, with the aim of using the responses from this sample to make some general statement about the whole of London. That is, using the sample data, we would like to estimate unemployment in the whole of London.

Example 4: More abstractly, in Chapter 4 I introduced the notion of a 'parameter' of a distribution. We saw the example of the Bernoulli family of distributions, where a random variable could take values 0 or 1, and where p was a parameter giving the probability of observing a 1. We also saw the example of a normal distribution, which had two parameters, its mean and standard deviation. Our aim might be to estimate the value of such a parameter. For example, an anthropologist might be studying the heights of a particular group of people. She might be prepared to assume that the heights were normally distributed, but to characterize the distribution fully she will need to know the mean and standard deviation of this distribution. She would like to use the heights of a sample of people from the group to estimate the mean and standard deviation of the entire population.

Point estimation

A friend offers me the following deal. He will repeatedly toss a coin, and whenever it comes up heads he will give me £10, but whenever it comes up tails I will give him £5.

At first glance, this looks like a good deal for me. After all, it is well known that coins are equally likely to come up heads and tails (probability of heads equals 1/2), so I am just as likely to win £10 as lose £5 on each toss. On average, I will be a winner.

But then I become suspicious. Why would he be offering me a deal which appears to be so much in my favour? I begin to suspect that perhaps the coin has been tampered with, so that the probability that it will show heads is in fact less than a half. After all, if the probability of it showing heads is really very small, so that it rarely comes up heads, the deal could be a poor one for me. To sort this out, what I would like is an estimate of this probability. My friend, who is very obliging but knows no statistics, offers to flip the coin six times, so I can see how it falls on each of those times. My aim, then, is to use these data to estimate the probability that the coin will come up heads in future tosses.

Suppose that the coin *had* been tampered with, and that the probability of it showing heads on any one toss was only 1/3. Since tosses of the coin are independent (the outcome of one toss does not affect the outcome on any other), we know that the probability of getting heads in two tosses is simply the product of the probability of getting heads on each toss: $1/3 \times 1/3 = 1/9$. Similarly, since the probability of a tail is $1 - 1/3 = 2/3$, the probability of getting a head followed by a tail would be the product of 1/3 and 2/3, that is 2/9. In general, assuming that the probability of getting a head on each toss is 1/3, we can calculate the probability of getting any sequence of heads and tails – and, in particular, a sequence identical to that observed in the six tosses we actually saw. For example, if the six tosses showed HTHTTT, the probability of obtaining an identical sequence by chance would be $1/3 \times 2/3 \times 1/3 \times 2/3 \times 2/3 \times 2/3 = 16/729$, which is approximately 0.022.

In the same way, we can calculate the probability of getting the HTHTTT sequence if the probability of heads on each toss really had any other value. For example, if the probability of heads is 1/2 (so the probability of tails is $1 - 1/2 = 1/2$), the probability of obtaining such a sequence is 1/2 × 1/2 × 1/2 × 1/2 × 1/2 × 1/2 = 1/64, which is approximately 0.016. And if the probability of heads is 1/10, the probability of obtaining such a sequence is approximately 0.007. And so on.

Now, our aim is to estimate the probability that the coin will come up heads in any future toss. That is, we want to pick a single value – 1/3 or 1/2 or 1/10, or whatever – as an estimate of this probability. Looking at the calculations above, we see that the probability of obtaining the observed outcome for the six tosses is 0.022 if the true probability of heads is 1/3, whereas it is only 0.016 if the true probability of heads is really 1/2, and it is lower still, only 0.007 if the true probability of heads is really 1/10. What this means is that we are more likely to get the observed six tosses if the true probability is 1/3 than if it is 1/2 or 1/10. It thus seems sensible to pick the value of 1/3 as our single estimate of the probability that heads will show. This is the value most likely to yield the data we actually obtained.

This example illustrates the *maximum likelihood* approach to estimation: we choose that value of the parameter which has the highest probability of yielding the observed data. In the example, I only calculated this probability for three values of the probability of heads coming up (1/3, 1/2, 1/10), but in principle we could calculate it for all possible values. The function showing the probability of the observed data for each possible choice of the probability of heads is called the *likelihood function*. This function plays a central role in statistical inference.

The same sort of principle can be applied to obtain estimates of the parameters of the normal distribution, or any other distribution. For different choices for the possible values of the parameter, we simply calculate what would be the probability of obtaining a data set like that actually obtained. Then the maximum likelihood estimator is that parameter value which yields the greatest probability. Note that this procedure yields a single value, an estimate which is best in the maximum likelihood sense. Because it is just a single value, it is called a *point estimate.*

An alternative way of thinking about this approach to estimation is to regard the likelihood function as a measure of agreement between the observed data (our sequence of six coin toss results) and what our theory predicts (where 'theory' here means a suggested value for the probability of being heads; for example, 1/3 or 1/2). Choosing the theory (the probability of getting heads) to maximize agreement, or, equivalently, to minimize discrepancy, is clearly sensible. Thinking of it in this way allows us to generalize: we can consider other measures of discrepancy. For example, in many situations a good measure of discrepancy is the sum of squared differences between the proposed parameter value and individual sample values. Choosing the parameter to minimize this measure means that a 'best' estimate is obtained, in the sense of smallest sum-of-squared-differences. In fact, this is a very common approach to estimation. It is called, for obvious reasons, *least squares estimation.*

Sometimes we might have ideas, before analysing the data, of the sort of value we expect the parameter to have. Such ideas might have come from previous experience or earlier experiments. For example, based on our previous experience in tossing coins, we might believe that the parameter p, giving the probability that a tossed coin will show heads, is near to 1/2, and that it is very unlikely to be far from 1/2. We say that we have a *prior distribution* of our belief that the unknown parameter takes different values. This distribution represents a subjective belief about the value of the parameter – as with the subjective interpretation of probability discussed in Chapter 4. In such cases, rather than analysing the data in isolation to yield an estimate for the value of the parameter, it makes sense to combine the data with our prior belief to yield a *posterior distribution* of our beliefs about the likely values of the parameter. That is, we start out with a distribution describing our beliefs about the possible values of the parameter, and we adjust this according to what we observe in the data. For example, our prior distribution for the probability that a coin will come up heads might be heavily concentrated around the value of 1/2: we think it is highly likely to be near 1/2. However, if 100 coin tosses show heads only 3 out of the 100 times, we might want to adjust that distribution, so that smaller values of the probability are regarded as more likely and values near 1/2 less likely.

In fact, it is Bayes's theorem, described in Chapter 4, which enables us to combine prior beliefs with observed data to give posterior beliefs. For this reason, this approach to estimation is termed the *Bayesian* approach. Recall that Bayes's theorem relates two conditional probabilities: the probability of A happening given that B has occurred, and the probability of B happening given that A has occurred. In the present case, we use the theorem to relate the probability that the parameter has some value, given the data we observe, to the probability of observing such data, given a particular value of the parameter. Now, the second of these, the probability of observing such data given a particular value of the parameter, is just the likelihood function. Bayes's theorem thus uses the likelihood of the data to adjust our prior beliefs, to yield our posterior beliefs.

Note that there is a subtle but important difference between this approach and the other approaches described above (often termed *frequentist* or *classical* approaches). There we assumed that the unknown parameter had some fixed but unknown value. For the Bayesian approach, however, we have assumed that the unknown parameter has a distribution over a set of possible values, initially given by the prior distribution, and then, when updated by the information in the data, by the posterior distribution. The researcher is acknowledging that the parameter could have different values, and using the probability distribution to express their belief about each value.

The notion of a prior distribution is not without its controversial aspects. At the very least, different people, with different background experience, might be expected to have different prior distributions. These would be combined with the data to yield different posterior distributions, and possibly different conclusions. Any pretence to objectivity has thus been sacrificed. There is also a practical difficulty. While the mean of a normal distribution and the parameter p in a Bernoulli distribution have clear and straightforward interpretations, it is not always the case that the parameters of distributions have straightforward interpretations. It can sometimes be very difficult coming up with sensible prior distributions reflecting our prior knowledge.

At this point in our description of the Bayesian approach we have arrived at the posterior distribution, a distribution summarizing the researcher's belief that the parameter takes each value, after having seen the data. If we wish, we can reduce that entire distribution to a single point estimate by using some summary statistic of the distribution. For example, we could use its mean or its mode.

Which estimate is best?

How can we tell if a method of point estimation is effective, and which of several estimators is best? For example, while I might choose to estimate the mean of a distribution using the mean of a sample drawn from that distribution, an alternative would be to drop the largest and smallest values of the sample before calculating the mean. In general, the largest and smallest values have greatest variability from sample to sample, so perhaps a more reliable and less variable estimate would result from dropping them.

For the frequentist approach to estimation, which assumes that there is some fixed, but unknown, true value for the parameter being estimated, we would ideally like to know which of these two approaches yields an estimate closer to the true value. Unfortunately, since the true value is unknown (the whole point is to estimate it!), we can never know this. On the other hand, what we *can* hope to know is how often we might expect the estimated value to be close to the true value if we were to repeat the exercise of taking a sample of measurements and calculating an estimate. After all, since the estimated value is based on a sample, it is likely that the estimated value would be different if a different sample was drawn. This means that the estimate is itself a random variable, varying from sample to sample. As a random variable, it has a distribution. If we know that this distribution is tightly clustered about the true value, we might regard the estimation method as a good one. Put another way, if we knew that a method *usually* yielded an estimate which was very near to the true value of a parameter, we might regard that as a good method of estimation. Whilst this tells us nothing about our particular case, we would justifiably have confidence in the method. After all, if you knew that 999 out of 1000 times someone made a correct prediction, you would surely be inclined to trust them in any particular case. You do this with train drivers, pilots, restaurants, etc: you know that the driver and pilot rarely crash, and the restaurant rarely serves contaminated food, so you are happy to take the risk that *this time* things will be OK.

Using this principle, several different measures have been developed to evaluate alternative frequentist estimation methods. One such measure is *bias*. This tells us how large the difference is between the true value of a parameter and the mean value of the distribution of estimated values. In particular, if this difference is zero (that is, if the mean of the distribution of estimated values is equal to the true value) then the estimator is said to be *unbiased*.

For example, the proportion of heads obtained when a coin is tossed several times is an unbiased estimator of the probability that the coin will come up heads: the mean value of the distribution of this proportion in repeated experiments is equal to the true probability that it will come up heads. To illustrate, suppose that, unknown to us, the true probability that a coin will come up heads is 0.55. We toss a coin ten times, and estimate this probability by the proportion of heads. Our ten tosses might yield six heads; that is a proportion of 0.6. Or three heads; a proportion of 0.3. Or five heads; a proportion of 0.5. And so on. On average (averaged over imaginary repetitions of the ten tosses) the proportion will be 0.55 because the proportion of heads is an unbiased estimator of the probability that the coin will show heads.

In general, an estimator which has a large bias will not be regarded as favourably as one which is unbiased. On average, over repetitions of the experiment, an estimator with large bias would yield a value very different from the truth.

The *mean squared error* is another measure of how good an estimator is. For any particular estimated value we could, if we knew the true parameter value, calculate the squared difference (the 'squared error') between the estimate and the true value. Squaring is useful, for one reason because it makes everything positive. Now, since the estimate itself is a random variable, varying from sample to sample, so also is this squared error. As a random variable, it has a distribution. The *mean* squared error is simply the mean of this distribution. A small mean squared error means that, on average, the squared difference between the estimated value and the true value is small. An estimator which is known to have a large mean squared error would not be regarded as favourably as one which had a small mean squared error: one would not have much confidence that its value was near to the truth.

Interval estimation

When we considered some basic summary statistics in Chapter 2, we saw that it was all very well summarizing a sample of values by their mean or some other single summary, but that this left a lot to be desired. In particular, it failed to show how widely the sample values were spread about this mean. We tackled that problem by introducing further summary statistics, such as the range and standard deviation, which indicated how widely dispersed the sample values were.

The same sort of principle applies in estimation. So far we have looked at point estimates, that is estimates which are *single* best estimated values in some sense. An alternative is to give a range of values, an *interval*, which we are confident includes the true value. Let us return to the £10/£5 deal offered by my friend. Previously we sought the single best estimate for the probability that a toss of the coin would produce heads. Instead, we could seek a range of values which we are confident will include the true probability. Perhaps we can be very confident that the true probability lies between 1/4 and 2/5, for example. This is an example of an *interval estimate*.

Now, since the true value is unknown, we cannot say for certain whether any particular interval will actually include the true value. But imagine repeating the exercise again and again with different random samples (just as we imagined when we defined bias above). For each of these samples we could calculate an interval estimate. Then, if the intervals are constructed in the right way it *is* possible to say that a certain percentage of the intervals (e.g. 95% or 99% or whatever we choose) would include the unknown true value.

Returning to my friend's coin, we cannot say for certain that any particular interval, calculated for any particular data sample, will contain the true probability that the coin will show heads. But we can say that 95% (or whatever we choose) of such intervals will contain the true probability. Since 95% of such intervals will contain the true value, we can have considerable confidence that the one interval we did calculate, based on the sample we actually obtained (HTHTTT in the example) would include the true value. For this reason, such intervals are called *confidence intervals*.

Turning to Bayesian methods, we saw that the outcome of a Bayesian analysis is an entire posterior distribution of values. This distribution tells us the strength of our belief that the parameter has any particular value. We could leave things at that. For example, if the distribution had a small standard deviation it would mean we were very confident that the parameter value lay in a narrow range. But sometimes it is convenient to summarize things in a way rather analogous to the confidence intervals above, and give an interval, defined by a largest and smallest value. For example, we could find an interval which contained 95% of the area beneath the posterior probability distribution within it. Since the distributions have the degree of belief interpretation, such intervals can be interpreted as giving the probability that the true value lies within them. To distinguish them from the frequentist confidence intervals, such intervals are called *credibility intervals*.

Testing

Statisticians use the phrases *hypothesis testing* and *significance testing* to describe the processes of exploring whether parameters in a model take specified values or lie in certain ranges. At its simplest level, this might mean testing just a single parameter. For example, we might know that 50% of patients suffering from a particular disease recover under the standard treatment, and we might conjecture that a proposed new drug treatment cures 80% of such patients. The single parameter we are interested in testing is the cure rate of the new treatment, and we would like to know if it is 80% rather than 50%.

Now, it is a fact that people are different. They differ in terms of age, sex, fitness, severity of disease, weight, and a host of other things. This means that, when even similar people are given the same dose of the same drug, the responses differ: some may be cured and some not. Indeed, it is entirely possible that the response will differ for the same patient at different times and under different circumstances. A reasonable model for this situation might be that a patient given a drug has a probability p of being cured. In our example, we know that $p = 0.5$ under the standard treatment and we conjecture that $p = 0.8$ under the new treatment.

In principle, at this point, to find what proportion are cured by the new drug, what we would like to do is give the new drug to everyone in the patient population, under all possible circumstances, and see what proportion are cured. This is clearly impossible, and what we have to do is give the drug to just a sample of patients. We can then calculate the proportion cured in the sample. Unfortunately, since we are merely working with a sample, and not the entire population, the mere fact that, say, 80% of the sample is cured, or 60%, or 90%, or whatever, does not necessarily mean that that proportion would be cured in the population. If we drew a different sample, we would be likely to obtain a different result.

However, a sample drawn from a population in which, overall, only 50% of the patients are cured will usually have a lower proportion cured than a sample drawn from a population in which 80% of the patients are cured.

This means that we can adopt a threshold, t say, such that if we observe the proportion cured in the sample to be less than t we will favour the 50% hypothesis, and if we observe a sample proportion cured to be greater than t we will favour the 80% hypothesis. In the latter case, we say that the sample statistic lies in the *rejection* or *critical region*, since the cure rate of the standard treatment, 50%, has been 'rejected'.

In doing this, we risk making one of two kinds of mistake. We might decide that the new drug cures 80% of the patients in the overall population when in truth it cures only 50%. Or we might decide that the new drug cures 50% of the patients in the overall population when in fact it cures 80%. The so-called *Neyman-Pearson* hypothesis testing approach arranges things so that the probability of making each of these two kinds of errors is known, and is sufficiently small to give us confidence in the conclusions.

Here is how it works. We begin by making a working assumption: let us assume that the new drug cures only 50% of patients. This working assumption is called the *null hypothesis*. The so-called *alternative hypothesis* is that the new drug cures 80% of the patients. Using basic probability calculations we can work out what proportion of samples would show a cure rate, by chance, greater than any chosen t, if the 50% assumption (the null hypothesis) were true. Typically, t is chosen so that, if the null hypothesis were true, only 5% or 1% of the time would the sample proportion cured exceed t.

In this situation, when the null hypothesis is true (i.e. if only 50% of the overall population would be cured) and we actually obtained a sample cure proportion greater than t, leading us to decide in favour of the overall 80% cure rate, we would be making the first kind of error noted above (which is conventionally called a *Type I error*). The symbol α is typically used to represent the probability of a Type I error. Our choice of t in the example means that we have fixed α at 0.05, or 0.01, or whatever value we chose.

If we observe a sample cure proportion greater than t, then either the null hypothesis is true (true rate of 50%) and an event of low probability (sample rate higher than t, occurring with probability α) has occurred, or the null hypothesis is incorrect. These are the only possibilities. This is the essence of the Neyman-Pearson approach to hypothesis testing. By choosing t so that α is small enough (and 0.05 and 0.01 are generally thought of as small enough), we feel reasonably confident in suggesting that the null hypothesis is not true because, if it was, an unlikely event would have occurred.

The other kind of error (*Type II*, naturally) arises when the alternative hypothesis is true (the 80% one in the example) but the observed sample cure proportion is less than t. Since we chose t to control the probability of making a Type I error, we cannot choose t also to control the probability of making a Type II error. However, we can make the probability of a Type II error as small as we like by taking a large enough sample. This is again a consequence of the law of large numbers. Increasing the sample size decreases the range of variability of the sample estimate, and hence decreases the probability that the sample estimate will be below t when the true population value is the higher, 80% value. In particular, by making the sample large enough we can reduce the probability of a Type II error to whatever value we think appropriate. The symbol β is typically used to represent the probability of a Type II error. The term *power* is used to represent $1 - \beta$, the probability of choosing the alternative hypothesis when it is true.

The hypothesis testing situation described above is analogous to the situation in a court of law, where the accused is initially presumed innocent (null hypothesis), and where two kinds of mistakes can arise: an innocent person is found guilty (Type I) or a guilty person is found innocent (Type II).

Note that two hypotheses are involved in Neyman-Pearson hypothesis testing: the null hypothesis and the alternative hypothesis. In *significance testing*, only the null hypothesis is considered. The aim is to 'reject' the null hypothesis if a value of some test statistic (the sample proportion cured in the example above) is sufficiently different from what would be expected under the null hypothesis, or 'fail to reject' it if the value is not so extreme. No alternative hypothesis is explicitly mentioned. The term *p-value* is used to describe the probability that we would observe a value of the test statistic as extreme or more extreme than that actually observed, if the null hypothesis were true.

The ideas of hypothesis and significance testing have been developed for a huge variety of problems. Particular tests have been developed often named after one of the original developers (e.g. the Wald test, the Mann-Whitney test) or named after the distribution of the test statistic involved (e.g. the t-test, the chi-squared test).

In principle, at least, Bayesian hypothesis testing is more straightforward. Under the Bayesian formulation, we have posterior probabilities that each hypothesis is true, so we can use these to choose a hypothesis. In practice, things are sometimes rather more complicated.

Decision theory

I informally described 'testing' as seeing if the parameters of a model took particular values or fell in certain ranges. This is a good description of much of what goes in a scientific context: the aim is to discover how things are. But in other contexts, such as commerce or medicine for example, the aim is typically not simply to discover what values the parameters have, but to act on this information. We want to look at a patient, make a number of observations and tests, and, using the resulting data, take the best course of action. 'Best' might mean many different things, but, speaking abstractly, we will want to maximize gain, profit, or 'utility', or, equivalently, to minimize cost or loss. If we can define a suitable such *utility function*, describing what the gain will be if each action is taken when the unknown truth takes each of its possible values, then we can compare different *decision rules* – that is, different ways of choosing between actions. For example, we might choose that decision rule which maximizes the minimum gain that could be incurred, whatever the unknown truth. Alternatively, if we are working within a Bayesian framework, and so have a posterior distribution of probabilities across the unknown state of the truth, we could calculate the average value of the gain for each decision rule, and choose that which had the largest average value.

Here is an example. A company might want to know which course of action, sending a letter or making a phone call, is most effective in encouraging its customers to buy its latest product. Now, it would be unrealistic to imagine that the same action would be most effective for all kinds of customers. Some will respond better to the letter, some to the phone call, and we do not know which is which. But the company might have data about each customer: the information they supplied when they first enrolled, the data describing their previous purchases, and so on. Using these data, we can formulate decision rules which say things such as 'if the customer is aged less than 25 and has a previous pattern of regular purchases then take action "phone call"; otherwise take action "letter"'. Many such potential decision rules could be formulated. For each of the actions, phone call or letter, we could estimate the gain, perhaps even in monetary terms, if we took that action and the customer turned out to be the type who did (or did not) respond well to that action. And then we could choose the decision rule which made the minimum gain the largest. Or we could average over the distribution of customers of each type, to yield an average gain for each decision rule, and then choose that rule which led to the largest average gain.

So where are we now?

Over the years, statistical inference has been the subject of considerable controversy, sometimes quite heated. Although different approaches to inference do sometimes lead to different conclusions, experience shows that sensitive use by statisticians who understand the methods they are using generally leads to similar conclusions. This is all part of the art of statistics and shows that carrying out a statistical analysis is not merely a mechanical exercise in mathematics. It requires understanding of the data and their background, as well as a sound grasp of the underlying inferential theory.

Different schools of statistical inference place varying degrees of emphasis on a number of different principles. Examples of these principles are the *likelihood principle* (if two different models have the same likelihood function, then they should lead to the same conclusions), the *repeated sampling principle* (statistical procedures should be assessed based on how they would behave 'on average' if they were applied to many repeated samples), and the *sufficiency principle* (concerned with summarizing data so that information sufficient for estimating a parameter is retained). Each of these principles seems perfectly reasonable, but they may sometimes conflict.

For many years the classical frequentist methods were the most widely used methods of inference, but Bayesian methods have gained considerably in popularity in recent years. This has been as a direct consequence of the development of powerful computers and clever computing methods, as well as of enthusiastic promotion of such methods by their supporters. Science takes place in a social context, and the human aspects of how different ideas about inference have gained and waned in ascendancy over the past few decades is a fascinating story.

One final point: as I hope has been made apparent in this chapter, there are different aspects to inference. In particular, we may be interested in trying to find answers to different kinds of questions. These include: what do the data tell me, what should I believe, what should I do, and so on. Different approaches to inference are best suited to different kinds of questions.

Chapter 6
Statistical models and methods

> The best thing about being a statistician is that you get to play in everyone's backyard.
>
> John W. Tukey

Statistical models: putting the blocks together

I have used the phrase 'statistical model' at various places in this book without so far defining what I mean. A statistical model is some simple representation or description of the thing or system being studied. A very simple model might involve just one aspect of nature. Indeed, we saw examples of this in Chapter 4 when we looked at distributions of single variables. More generally, statistical models can be very elaborate indeed, perhaps involving thousands of variables related in highly complicated ways. Economists trying to guide the decisions of a national bank will use such large models, for example.

A basic perspective on models is to ask whether they properly represent the underlying reality: whether they are 'true' or not. Indeed, this is the perspective we took earlier in the book, when we asked if a proposed parameter value was the true value. However, a more sophisticated perspective acknowledges that no model, statistical or otherwise, can take into account all of the possible influences and relationships in the real world. It is this sort of perspective which has led the eminent statistician George Box to assert that 'all models are wrong, some models are useful'. We build models for a reason: to help us understand, predict, decide, and so on. And while we recognize that our models represent a necessary simplification of the awesome complexity of the world, if we choose them well then they will enable us to do these things. But if we choose them badly, then we will not understand, our predictions will go awry, and our decisions will lead to mistakes. Our aim, then, is to construct models which are good enough for our purpose.

Statistical models may be conveniently divided into two types, often called *mechanistic* and *empirical* models. A mechanistic model is based on some solid underlying theory for how things are related. For example, a theory in physics might tell us how the speed of falling objects increases with the time for which they fall. Or another theory might tell us how drugs will disperse throughout the body. In both of these cases, the models will be based on theories about how things actually work. Indeed, the models will be based on the mathematical equations describing these theories, and the data we collect to evaluate our models will be values of the variables used in the theories, such as speed and time (in the falling object case) and concentration and time (in the drug diffusion case). Mechanistic models are thus direct mathematical ways of describing theories.

In contrast, empirical models are simply attempts to provide convenient summaries for the important aspects of observed data. We might have no theory which says that falling objects increase their speed as time passes, but we may observe a relationship between time and speed and, on the basis of this, conjecture some increasing relationship. If there is no underlying theoretical basis for this proposed relationship, the model would be an empirical model.

Mechanistic models are widespread in the physical sciences and disciplines such as engineering. The social and behavioural sciences tend to make more use of empirical models. Having said that, obviously there is considerable overlap: the nature of the model will depend on what is being modelled and how well it is understood. Economics, a particular social science, is full of mechanistic models based on theories about how economic factors are related. In general, it is probably fair to say that, in the initial stages of exploration of a phenomenon, empirical models are more common since regularities and patterns are being sought in the mass of observations. In later stages, when understanding has grown, so mechanistic models become more important. In any case, as our models for falling objects show, a particular model can be constructed as empirical and then become mechanistic, as understanding of the phenomenon grows.

Sometimes it is useful to distinguish between the various possible uses of statistical models. One such distinction is between *exploration* and *confirmation*. In exploration, we seek relationships or patterns. In confirmation, we aim to see if data support a proposed explanation. So, for example, in an exploratory study we might look for variables that are closely related. Perhaps one variable takes a high value whenever another one does, or perhaps sets of variables take very similar values for different objects, and so on. In confirmatory studies, on the other hand, we might use the data to estimate the parameters of a proposed statistical model and carry out a statistical test to see if the estimate is close enough to what our theory predicted. Statistical methods of data exploration have become increasingly important in recent years, with larger and larger data sets accumulating. This is true for both scientific applications (e.g. particle physics and astronomy) and commercial applications (e.g. databases containing details of supermarket purchases, telephone calls, or internet click stream data).

Another important distinction in statistical modelling is between *description* and *prediction*. In describing a data set, the aim is to summarize it in a convenient way. For example, if the data set consists of observations of ten variables (height, weight, time to travel to work, etc.) on each of a million people, then to begin to understand it we need to reduce it to a manageable size. For example, we could summarize it in terms of the means and standard deviations of each of the variables, as well as measures of how closely they were related. Then we would have some hope of understanding what is going on since we would have described the general properties of the data in a convenient way. Having said that, as we saw in Chapter 2, such descriptive summaries are not without their risks. By definition, they simplify the immense complexity of the entire data set, so we must be alert for the possibility that our summary description has left out something important. For example, perhaps our model has failed to take account of the fact that there are two distinct genetic groups in a population, so that a more elaborate model is needed to represent this.

In prediction, our aim is to use some of the variables to predict values of others. For example, we might have a collection of data showing details of childhood diet and their later adult height for a sample of people. Using this, we could construct a model relating adult height to childhood diet, and then use the model to predict the likely future height of a child following a particular diet. Note a fundamental aspect of the data needed for such modelling: we need values for both the predictor variables and the predicted variable from our sample. This will turn out to be a very important distinction between predictive and descriptive models, as we will see below.

Once again, the distinction is not always clear cut. We might simply be concerned with describing the relationship between childhood diet and adult height, with no intention to use the model to predict one from the other.

Another important kind of prediction is *forecasting*. Here we use data from the past to construct a model which can be used as the basis for predicting likely values of observations yet to be made. For example, we might look at the monthly pattern of sales of television sets over the past five years and extrapolate the trend in sales and the seasonal variation to forecast the likely sales over the next twelve months.

Statistical models also have other uses. We briefly saw their role in decision making in Chapter 5. We also saw in Chapter 5 how the parameters of distributions were estimated. This is done by defining a measure of discrepancy between the observed data and the theoretical distribution, and then choosing the estimated parameter value which minimizes the discrepancy measure. A common measure of discrepancy was derived from the likelihood, measuring how probable it was that data like the observed data would arise if the parameters took various different values. Now, since distributions are merely simple forms of model, exactly the same principles apply when fitting more elaborate models (such as those illustrated below). However, a curious phenomenon arises as the models become more and more elaborate.

I shall take a simple example to illustrate. Suppose we want to construct a model to predict initial salaries of graduates, based on data describing their schooling, the subjects they studied at university, their examination scores, and also factors such as age, sex, where they lived, and so on. Suppose we sample 100 new graduates and collect the data from them. Now, in general, if we try to base our prediction on very few variables (e.g. just age) then we will not obtain very accurate predictions. Age, by itself, just does not contain enough information to allow us to say very precisely what someone's graduate salary will be. To improve the predictive accuracy we need to add more predictors (e.g. use age *and* subject of study *and* exam scores to predict graduate salary). However, and here comes the crunch, if we add too many predictor variables then the predictive accuracy for the population decreases. Even though we are making use of more information about the graduates, our model is not as good.

This seems counterintuitive. How can adding *more* information lead to *worse* predictions?

The answer is subtle, and goes under various names, including the graphic *overfitting*. To understand it, let us take a step back and see what our real aim is. Our aim is *not* to get the best predictions we can for the 100 graduates in our sample: we already know their initial salaries. Rather, it is to get the best predictions we can for *other* graduates. That is, our aim is to *generalize* from the sample we have. Now, by adding more and more predictor variables we are certainly adding information which will enable us to predict more and more accurately the salaries of those already in our sample. But the sample is only a sample: it does not fully represent the salaries of the entire population of graduates. And, after a while, as we continue to add more predictor variables, so we start to predict aspects of the data which are peculiar to the sample. They are not features which apply to the more general population.

This phenomenon applies to all statistical modelling: models can be too complicated, so that they fit the observed data very well indeed but fail to generalize well to other objects drawn from the same distribution. It means that it is necessary to develop strategies for choosing models of the right complexity: too simple and we risk missing out on potential predictability, too complex and we risk overfitting. This principle underlies Occam's razor, which states that 'models should be no more complicated than is necessary' (attributed to the 14th-century Franciscan friar William of Occam).

The overfitting problem is particularly important in modern statistics. Prior to the advent of the computer, and before it became commonplace to fit complicated models with very large numbers of parameters, there was less risk of overfitting.

Statistical methods: statistics in action

The aim of this section is to outline some important classes of statistical method, to show how they are related, and to illustrate the sorts of problems they can be used to solve.

Let us begin by noting that we are frequently interested in relationships between pairs of variables. Does risk of heart attack increase with body mass index? Is global warming a consequence of human activity? If unemployment goes up will inflation go down? Will improving a car's safety features increase its sales? And so on. If two variables are related in the sense that larger values of one tend to be associated with larger values of the other, then the variables are said to be *positively correlated*. If larger values of one tend to be associated with smaller values of the other, they are said to be *negatively correlated*. Height and weight in humans are positively correlated: taller people tend to be heavier. Note that the relationship is not an exact one: there are light tall people (the thin ones) and heavy short people. But, on average, overall, tallness is associated with greater weight. We can also see from this example that just because two variables are correlated does not mean that one causes the other. Putting someone on a diet of cream buns to increase their weight is unlikely to lead to an increase in height, and putting them on a rack to stretch them is unlikely to increase their weight. In fact, confusion between correlation and causation has been the source of much misunderstanding over the years. A random sample of children aged between 5 and 16 years old is likely to show a marked positive correlation between ability to read and ability to do arithmetic. But one is unlikely to cause the other. It is more likely that age is a common cause of each: the older children are better at both reading and arithmetic.

A single number, a *correlation coefficient*, can be used to represent the strength of a correlation. There are various ways in which this strength may be measured, just as we saw that there were various ways of defining 'average' and 'dispersion'. In general, however, correlation coefficients are standardized to lie between -1 and $+1$, with 0 meaning no relationship, $+1$ meaning a perfect positive correlation, and -1 meaning a perfect negative correlation. A 'perfect' correlation between two variables x and y means that if you know x then you know y exactly.

Correlation is a symmetric relationship: if height is correlated with weight, then weight is correlated with height, and the strength of this correlation is the same whichever way we look at it. In contrast, sometimes we are interested in asymmetric relationships between variables. For example, we might want to know how much weight difference, on average, is associated with a height difference of ten centimetres. This sort of question is answered by the statistical technique of *regression analysis*. A regression model tells us what is the average value of a variable y for each value of a variable x. In the example above, a 'regression of weight on height' would tell us the average weight that people of each height would take. This is illustrated in Figure 5, where weight is plotted on the vertical axis, and height on the horizontal axis. Each black dot shows the (weight, height) pair for a person from our sample. Now it is obvious from this figure that we do not have observed values for *all* possible heights. For example, there is no data point with a height of exactly 6′. One way to overcome this difficulty, to construct a model which gives us an average weight for each value of height, is to suppose that there is a simple relationship between height and average weight. A very simple such relationship is a straight line relationship; an example of such a line is shown in the figure. For any given height, this line allows us to look up the corresponding value of average weight. In particular, for example, it gives us a value for the average weight of people who are 6′ tall.

There are several points to make about this approach.

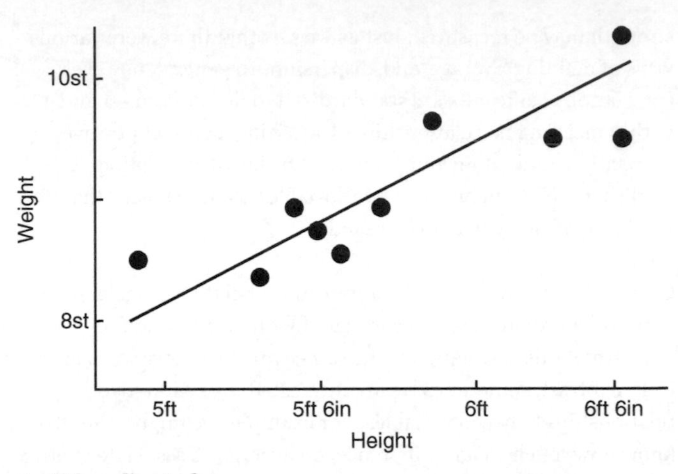

5. Fitting a line to data

First, it gives the *average* weights at each height. This is reasonable: in real life, even people of the same height can weigh different amounts.

Second, we need to find some way of determining exactly which line we are talking about. The figure shows one line, but how did we choose that line rather than some other? Now, lines are uniquely specified by two parameters, their intercept (here, the value of weight at which the line meets the weight axis) and slope, so we need to find some way of choosing, or estimating, these two parameters. But we know about parameter estimation; we studied it in Chapter 5. To estimate parameters we choose those values which minimize some measure of discrepancy between the model and the observed data. For any particular (weight, height) pair in the data, one measure of discrepancy is the squared difference (again, squared so as to make things positive) between the observed weight and the predicted weight at that height. An overall discrepancy measure based on this is the sum of squared differences between the observed weights and the predicted weights at the heights given in the data. We then estimate the intercept and slope by choosing those values which minimize this sum of squared differences. In the sense that it minimizes the (sum of squared) differences between the observed and predicted values of weight in the data, this *least squares regression line* yields the best prediction of average weight for any value of height we might care to choose.

The third point is that that the assumption of a straight line relationship might seem fairly arbitrary, adopted with little justification. Why choose a straight line, rather than a curved line? Without going into the details here, it is possible to introduce curvature of various degrees so that the line showing the relationship between height and average weight can have more complicated shapes – perhaps increasing more rapidly at lower heights than at higher heights, for example. We do this by making the model more complicated, by introducing extra parameters, in addition to the intercept and slope.

The height/weight regression example sought to predict average weight from only one predictor variable, height. We could also include other potential predictors, in order to yield more accurate predictions. For example, men and women have different body shapes, so that, for a given height, some of the difference in weights may well be due to gender. We might therefore also include gender as a predictor. We could continue this, including other variables we thought likely to be related to weight. We should not go too far if we have observations on only a fixed number of people, or once again our model will overfit the data. We therefore might not want to include all of the variables we can think of, but simply include a subset of them.

In general, there are also other reasons why we might want to include only a subset of the potential predictor variables. For example, perhaps measuring additional predictor variables is expensive, or takes a long time, so we will want to keep the number to a minimum. For these and other reasons, statisticians have developed methods for finding good subsets of variables, where 'good' means that they yield the best predictions.

Regression models relate an outcome or response variable to one or more predictor variables. This is an extremely common type of problem, and other statistical models have been developed to cope with similar situations which differ in some ways from the straightforward regression situation. In *survival analysis*, for example, the value of the response variable is known only for some of the cases, and its value for the other cases known only to exceed some value. This arises most commonly (though certainly not only) when the response variable is time duration. Thus, we might want to know how long a patient will survive (hence the name of the technique) or how long a component of a system will last before requiring replacement. Taking the first case to illustrate, our data set might show that one of the patients lived 5 months, another only 2 months, three others lived 11 months, and so on. However, for practical reasons perhaps we could not wait until the last patient in the study had died (which might be years hence), so we stopped taking observations. All we would then know about some of the patients is that they lived *longer* than the time between starting and stopping observations. Such data are described as *censored*. To illustrate the complications they introduce, consider the calculation of the average survival time. To calculate the average, we need to add up the observed times and divide by how many there are. Now we do not actually observe the survival times for those patients who have been censored, so we cannot include them in the calculation. But if we leave them out, we will be leaving out precisely those values which are largest, so our estimate will be biased downwards. Conversely, if we include them, using the observed durations, the result will depend on when we happened to choose to stop making our observations. Since this is equally inappropriate, more sophisticated methods have been developed which cope with censored data.

Another variant on the problem of having a single outcome variable related to one or more predictor variables occurs in *analysis of variance*. This is widely used in agriculture, psychology, industrial quality control, manufacturing, and other areas. In analysis of variance, the predictor variables are categorical, meaning that they each take only a few values. For example, in manufacturing some chemical we might be able to control temperature, pressure, and duration, and have three settings for each: low, intermediate, and high. This sort of situation arose when we discussed experimental design in Chapter 3, and analysis of variance is often used to analyse experiments. Although typically presented as rather different from regression analysis, it is possible to reformulate it as a regression model. Both are special cases of a broader class of model called a *linear model*.

Linear models themselves have been extended in various ways. One very important generalization is to so-called *generalized linear models*. In regression and analysis of variance, the aim is to predict the mean value of the response at each value of the predictor(s). Generalized linear models extend this by permitting other parameters of the distribution of the response, not merely its mean, to be the subject of the prediction.

Yet another variant of the outcome/predictor structure arises when the response is itself categorical. For example, the response might be a list of possible medical diagnoses, and the predictors might be a combination of symptoms (perhaps coded as present or absent) and the results of medical tests. Such methods go under the general name of *supervised classification*. The most important special case of such models arises when the response variable is binary, taking only two possible values, such as sick/healthy, good risk/bad risk, profitable/unprofitable, spoken word 'yes'/spoken word 'no' (in speech recognition), authorized fingerprint/unauthorized fingerprint (in biometrics recognition systems), fraudulent transaction/legitimate transaction, and so on. In each case, the aim will be to construct a model which will enable us to determine the most likely category of new cases, using only the information in the predictor variables.

A large number of statistical tools have been developed for such situations. Amongst the earliest was *linear discriminant analysis*, described in the 1930s but still very widely used today, both in its basic form and in more elaborate extensions. Another method which is very popular in some domains, such as medicine and customer value management, is *logistic discriminant analysis*. This is a variant of logistic regression, a type of generalized linear model, so showing the close link between the classes of tools. In fact, logistic regression can be regarded as the most basic kind of *neural network*. Neural networks are so called because they were originally suggested as models for the way the brain worked. Nowadays, however, the work in the area has largely focused on their statistical properties as prediction systems, regardless of whether or not they form good models of natural systems.

Other models for supervised classification include *tree classifiers* and *nearest neighbour* methods. A tree model splits variables into ranges, and classifies new points according to the combination of ranges in which they lie. For example, analysis of the data might show that people who are aged over 50, have a sedentary lifestyle, and have a body mass index greater than 25 are at risk of heart disease. Such models can be represented as tree structures; hence the name. In a nearest-neighbour method, we find the few objects in the data set which are most similar (or 'nearest') to the new object to be classified, where similarity is defined in terms of the predictor variables. Then the new object is simply assigned to the same class as the majority amongst these most similar objects.

Supervised classification is so called because it needs someone (a 'supervisor') to provide the class labels for a sample of data, from which we can construct the classification rule to apply to new objects. In other classification problems, however, there is no existing class label, and the aim is simply to divide up the objects into natural, or perhaps convenient classes. We might say that the aim is to define the classes. In medicine, for example, we might have a sample of patients for each of whom we have details of their symptom patterns and test results, and we might suspect that several distinct types of disease are represented in the sample. Our aim, then, would be to see if the patients form distinct groups, in terms of their symptoms and test results. Statistical tools for exploring such groupings are called *cluster analysis*. Such methods were helpful in identifying the distinction between unipolar and bipolar depression, and are used in a wide variety of other areas – including, for example, customer value management and marketing, where interest lies in deciding if there are different types of customer.

In cluster analysis, there is no 'outcome' or 'response' variable. Rather, the aim is simply to describe the data in a convenient way. Other statistical tools have the same objective, though the sort of description they seek is completely different. For example, a *graphical model* is a simplified description of the relationships between several, possibly a large number, of variables, based on the assumption that the relationships between many of the variables are caused by intermediate relationships with other variables. We saw a very simple example of this above: perhaps the positive correlation between reading ability and arithmetic ability of children was a consequence of the relationship between each of these variables and age.

Such models can be extended by supposing that some of the relationships are caused by unmeasured *latent* variables which are related to some of the observed variables and hence induce an apparent relationship between them. For example, we might observe that the stock market prices of certain companies increase or decrease together. One way to explain this might be to conjecture the existence of some unobserved variable (some aspect of the economy, for example), which is related to each of the prices, and which therefore induces the correlation between them: when the unobserved variable increases, so do all of the prices. Such ideas underlie *factor analysis* models: the latent variable is often called a latent *factor*. They also underlie *hidden Markov models*, in which a sequence of observed values is explained in terms of the hidden states of a system. For example, patients with some diseases fluctuate in quality of life, sometimes relapsing and sometimes making temporary recoveries. Such progression can be modelled in terms of changing underlying states.

If classification methods are named after the sorts of problems they are designed to solve, other methods are named after the nature of the data on which they work. *Time series analysis* methods, for example, work on time series: repeated observations of the same variable or variables, at a sequence of times. Such data structures are ubiquitous, occurring in economics (e.g. measurements of inflation, GDP, and unemployment), engineering, medicine (e.g. intensive care units), and any number of other domains. In analysing a time series, we might be aiming to understand it, to decompose it into key components (e.g. trend, seasonality), to detect when system behaviour changes, to detect anomalies (e.g. earthquake prediction), to forecast likely future values, or for a host of other reasons. A wide variety of methods have been developed for analysing such data.

Statistical graphics

One particular class of statistical tools is so important that it deserves special mention. This is the use of graphics. The human eye has been honed by aeons of evolution to be able to perceive structures and patterns in the signals reaching it. Statisticians make extensive use of this by representing data in a huge range of different kinds of graphical display. When data are displayed well, relationships between variables or configurations in data become obvious. This is used both in analysing data, to help understand

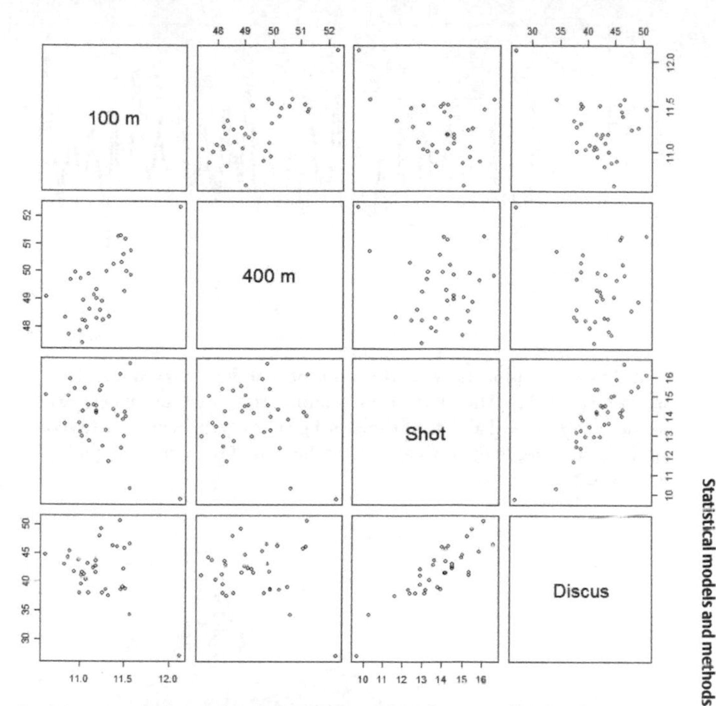

6. A 'scatterplot matrix', showing the times (in seconds) for the 100-metre and 400-metre sprint, and the distances (in metres) for the shot and discus for competitors in the men's decathlon in the 1988 Olympic Games. Each square shows the relationship between two of the four variables. The strong correlation between the scores in the two throwing events is immediately apparent

what is going on (recall the distribution of baseball salaries in Figure 1), and for communicating the findings to others. Some illustrations are given in Figures 6 to 8.

Conclusion

This chapter has presented a lightning review of just a few important statistical tools, but there are a great many others I have not mentioned. Different models are suited to different kinds

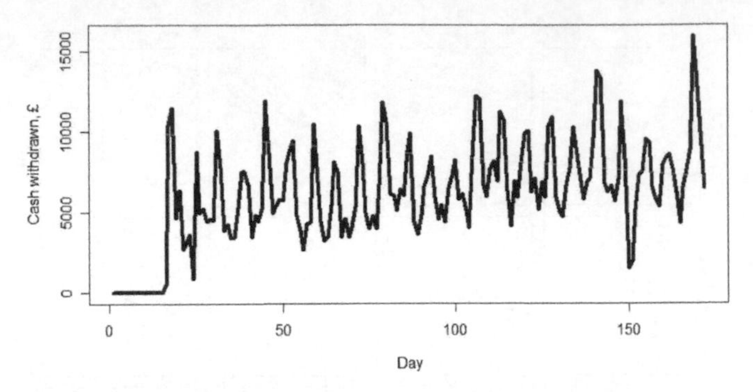

7. A time series plot showing the amount withdrawn from an ATM machine each day. The figure clearly shows that there are weekly and monthly cycles, and also that there is a gradually increasing trend over time. An anomalously low value near the end of the period is also apparent

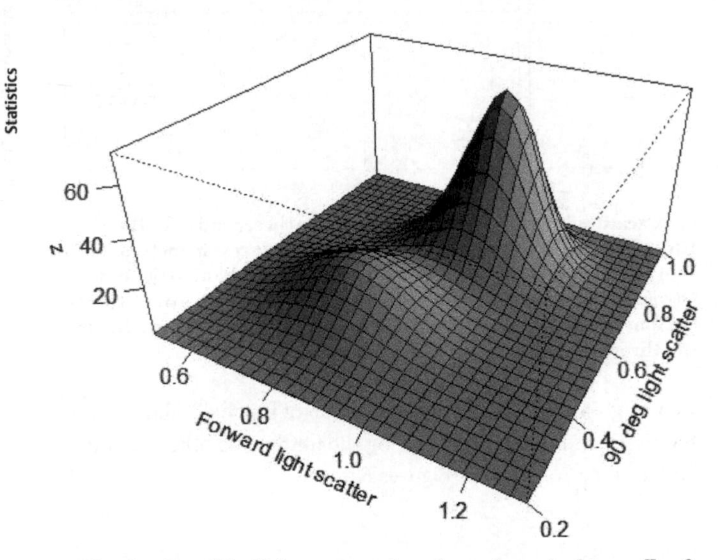

8. Distribution of the light scatter values from phytoplankton cells of different species. In fact, three species are shown here, but two of them have very similar distributions of values, so these combine to form a single high peak

of problems and different kinds of data, and there is an unlimited number of problems and data structures. It is also important to appreciate that models are not isolated entities. The truth is that different models are related in multiple ways. They may generalize or be special cases of other kinds of models or be adapted to different kinds of data, but they are all embedded in a rich network of relationships.

Chapter 7
Statistical computing

> The actual magic comes from our statistical analysis team.
>
> Sam Alkhalaf

Statistics changes its spots

In the discussions above we saw that overfitting could be a problem. We also left the solution rather in the air, simply saying that it was necessary to choose models which were neither too complicated nor too simple. Without substantial experience in statistical modelling that is not very helpful advice, and more objective approaches are needed. One is based on the principle of *cross-validation.*

We have seen that, in general, as the complexity of a model increases, so its goodness of fit to the available data continues to improve but that its goodness of fit to other samples drawn from the same distribution (or its 'out of sample performance') typically initially improves but then begins to deteriorate. Here the 'other samples' are representative of new data, which is what we are really interested in. The point at which the model best fits data from some 'other sample' would seem to give a model of the appropriate level of complexity. And that is the key to the solution: we should estimate the model's parameters using one sample, and evaluate its performance using some other sample.

Unfortunately, we typically have only one sample. One approach is therefore to (randomly) split this sample into two subsamples. One subsample (the *training* or *design sample*) is used for parameter estimation and the other (the *validation sample*) for assessing performance and choosing the model. This is the cross-validation approach. Typically, to ease any problems arising from the fact that the subsample used for estimating the parameters is not the entirety of the original sample, the procedure is repeated multiple times. That is, the original sample is randomly divided into two, parameters are estimated using one subsample, and the model is evaluated using the other. This is repeated for different random divisions of the sample. Finally, the evaluation results from each split are averaged, to yield an overall measure of likely future performance.

Cross-validation is an example of a *computationally intensive* approach – so called for the obvious reason that multiple models have to be built. Another important class of such methods is *bootstrap resampling*. Bootstrap methods have a variety of uses, but one important one is estimating the uncertainty associated with complex models; that is, determining how different we might expect the model to be if we had drawn a different sample of data. Bootstrap methods work by taking random subsamples of the same size as the original sample from the original sample (which means some data points will be used more than once). A new model, of the same form as that being evaluated, is built on each of these subsamples. It is as if we had multiple samples, all of the same size, from the original distribution, each yielding an estimated model. This collection of models can then be used to investigate how different the model would have been, had we drawn a different sample.

One of the most striking illustrations of how the power of the computer has changed modern statistics is in the impact of computer-intensive methods on the Bayesian approach to inference, described in Chapter 5. To use Bayesian methods in practice, it is necessary to calculate complicated functions of distributions (in mathematical terms, high-dimensional integrations are needed). The computer has allowed this problem to be sidestepped. Instead of evaluating the distributions mathematically, the computer draws large numbers of random samples from them. The properties of the distributions can be estimated from these random samples, in just the same way that we used the sample mean to estimate the mean of a population. Such *Markov chain Monte Carlo* methods have revolutionized the practice of Bayesian statistics, essentially transforming it from a theoretically attractive but practically limited set of ideas to a powerful technology for data analysis.

The previous chapter drew attention to the power of graphical methods, for both elucidation and communication, but the computer has shifted graphical methods to an altogether new plane. Whereas, in the past, we might have had static black and white images, we now have dynamic colour images. Even more importantly, we can now interact directly with the image. To take just one simple example, it is possible to simultaneously display multiple plots, each one showing the relationships between different pairs of variables associated with the objects, like the scatterplot matrix in Figure 6, but now with the displays linked via the computer. Then highlighting or otherwise manipulating a set of points manifests itself simultaneously in all the plots. Other tools allow one to dynamically 'fly' through high-dimensional data spaces, displaying the data in multiple ways.

Because statistics is used so universally, and because the computer plays such a central role, it is hardly surprising that user-friendly statistical software packages have been developed. Some of these are so important that they have become industry standards in certain application areas. But this should not lead us to forget that effective application of statistical tools requires careful thought. Indeed, in the early days of the development of statistical software, some feared that the availability of such tools would remove the need for the statistician, since then 'anyone could do a statistical analysis: all they had to do was give the computer appropriate instructions'. The fact is, however, that the reverse has proven to be the case. There is more and more demand for statisticians as time goes on. There are several reasons for this.

One reason is that, increasingly, data are recorded automatically. In everyday life, every time you make a credit card purchase or shop in a supermarket, details of the transaction are automatically stored; in the natural sciences, digital instruments record physical and chemical properties without needing human intervention; in hospitals, electronic devices automatically monitor patients; and so on. We are faced with a data avalanche. This represents a tremendous opportunity, but statistical skills are needed to take advantage of it.

A second reason is that new areas requiring statistical skills are appearing. Bioinformatics and genomics are teasing apart the awesome complexity of the human body from experimental and observational data, and are based on statistical inference. The hedge fund industry has been described as 'an industry built on statistics'. It uses statistical tools to model how stocks and other price indices behave.

A third reason is that it is one thing to give commands to a computer, but it is quite another to know what commands to give and to understand the results. It is certainly not merely a question of choosing the right tool for the job and letting the computer do the rest. It requires statistical expertise and understanding. For an amateur, it is important to know one's limits, and when one should call on the advice of an expert statistician. Regrettably, every week the media provide illustrations of people who are stretching themselves beyond their statistical understanding.

For these reasons and more, statistics is experiencing a golden age.

We have now reached the end of this very short introduction. We have seen something of the extraordinary breadth of statistics: the fact that it is applied in almost all walks of life. We have seen something of its methods: the sophisticated tools and procedures it uses. We have also seen that it is a dynamic discipline, still growing and developing. Above all, however, I hope I have made it clear that modern statistics, based on deep philosophical foundations, is the art of discovery. Modern statistics enables us to tease out the secrets of the universe around us. Modern statistics enables understanding.

Further reading

Chapter 1

A. R. Jadad and M. W. Enkin, *Randomised Controlled Trials: Questions, Answers and Musings*, 2nd edn. (Malden, Massachusetts: Blackwell Publishing, 2007).

Joel Best, *Damned Lies and Statistics: Untangling Numbers from the Media, Politicians, and Activists* (Berkeley: University of California Press, 2001).

John Chambers, Greater or lesser statistics: a choice for future research, *Statistics and Computing*, 3 (1993): 18–24.

Foundation for the Study of Infant Death. <http://www.fsid.org.uk/cot-death.html>. Accessed 6 April 2007.

Helen Joyce, Beyond reasonable doubt, *Plus Magazine* (2002). <http://www.plus.maths.org/issue21/features/clark/index.html>. Accessed 14 July 2008.

<http://www.sallyclark.org.uk/>. Accessed 14 July 2008.

Chapter 2

D. J. Hand, *Information Generation: How Data Rule Our World* (Oxford: Oneworld, 2007).

F. Daly, D. J. Hand, M. C. Jones, A. D. Lunn, and K. McConway, *Elements of Statistics* (Harlow: Addison-Wesley, 1995).

Chapter 3

S. Benvenga, Errors based on units of measure, *The Lancet*, 363 (2004): 1368.

T. L. Fine, *Theories of Probability: An Examination of Foundations* (New York: Academic Press, 1973).

Chapter 4

D. R. Cox, *Principles of Statstical Inference* (Cambridge: Cambridge University Press, 2006).
H. S. Migon and D. Gamerman, *Statistical Inference: An Integrated Approach* (London: Arnold, 1999).

Chapter 5

D. C. Montgomery, *Design and Analysis of Experiments* (New York: John Wiley and Sons, 2004).
L. Kish, *Survey Sampling* (New York: John Wiley and Sons, 1995).

Chapter 6

G. E. P. Box, Robustness in the strategy of scientific model building, technical report, Madison Mathematics Research Center, Wisconsin University, 1979.
E. Tufte, *The Visual Display of Quantitative Information* (Cheshire, CT: Graphics Press, 2001).
A. Unwin, M. Theus, and H. Hofmann, *Graphics of Large Data Sets: Visualising a Million* (New York: Springer-Verlag, 2006).

Endnote

In Chapter 1, answers to elementary misunderstandings:

(1) Clearly, the sooner a disease is detected, the longer that patient will still have to live, regardless of any medical intervention. Somehow this needs to be taken into account.

(2) A 25% reduction means the price is reduced by a quarter. But that means that to get back to the original price you have to increase the reduced price by a third (33%), not a quarter (25%). For example, a 25% discount on an original price of £100 leads to a stated price of £75. To get back to the original price we have to increase this by £25, which is 33% of £75.

(3) This assumes that life expectancy will continue to increase at the same rate as it has increased in the past.

(4) If one child was gunned down in 1950, the statement would mean that two were gunned down in 1951, four in 1952, eight in 1953, sixteen in 1954, and so on. Continuing to double in this way would mean that by now more children are gunned down each year than there are people in the world. (This example is from the excellent book by Joel Best, listed in the Further reading.)

Index

A

B

C

D

E

F

G

H

I

J

K

L

M

U

V

W